grow a *sustainable* diet

grow a
sustainable
diet

planning and growing to feed
ourselves and the earth

Cindy Conner

Illustrations by Betsy Trice

new society
PUBLISHERS

Today, more than ever before, our society is seeking ways to live more conscientiously. To help bring you the very best inspiration and information about greener, more sustainable lifestyles, *Mother Earth News* is recommending select books from New Society Publishers. For more than 30 years, *Mother Earth News* has been North America's "Original Guide to Living Wisely," creating books and magazines for people with a passion for self-reliance and a desire to live in harmony with nature. Across the countryside and in our cities, New Society Publishers and *Mother Earth News* are leading the way to a wiser, more sustainable world.

For more information, please visit MotherEarthNews.com.

Cover design by Diane McIntosh.
Gardening tools image © iStock (Mark Swallow); plate/table © iStock (sorendis);
garden plots © iStock (Skystorm); all interior illustrations by Betsy Trice.

Printed in Canada. First printing January 2014.

New Society Publishers acknowledges the financial support of the Government of
Canada through the Canada Book Fund (CBF) for our publishing activities.

Paperback ISBN: 978-0-86571-756-5 / eISBN: 978-1-55092-553-1

Inquiries regarding requests to reprint all or part of *Grow a Sustainable Diet*
should be addressed to New Society Publishers at the address below.

To order directly from the publishers, please call toll-free
(North America) 1-800-567-6772, or order online at www.newsociety.com

Any other inquiries can be directed by mail to:

New Society Publishers
P.O. Box 189, Gabriola Island, BC V0R 1X0, Canada
(250) 247-9737

LIBRARY AND ARCHIVES CANADA CATALOGUING IN PUBLICATION

Conner, Cindy, author
Grow a sustainable diet : planning and growing to feed ourselves
and the earth / Cindy Conner ; illustrations by Betsy Trice.

Includes bibliographical references and index.
Issued in print and electronic formats.
ISBN 978-0-86571-756-5 (pbk.).—ISBN 978-1-55092-554-8 (ebook)

1. Permaculture. 2. Organic gardening. 3. Gardens—Planning.
4. Gardening—Environmental aspects. I. Title.

S494.5.P47C65 2014   631.5'8   C2013-907439-2
                                                 C2013-907440-6

New Society Publishers' mission is to publish books that contribute in fundamental
ways to building an ecologically sustainable and just society, and to do so with the
least possible impact on the environment, in a manner that models this vision. We
are committed to doing this not just through education, but through action. The
interior pages of our bound books are printed on Forest Stewardship Council®-
registered acid-free paper that is **100% post-consumer recycled** (100% old growth
forest-free), processed chlorine-free, and printed with vegetable-based, low-VOC
inks, with covers produced using FSC®-registered stock. New Society also works to
reduce its carbon footprint, and purchases carbon offsets based on an annual audit
to ensure a carbon neutral footprint. For further information, or to browse our
full list of books and purchase securely, visit our website at: www.newsociety.com

Contents

Foreword

by John Jeavons

Grow A Sustainable Diet is just the book you have been looking for!

We are living in exciting times. There are many concerns about health-of both the human population and the planet. Those concerns open the door to opportunities for each of us to make a difference—and we can begin in our gardens at home! When we choose to eat food grown in a way that increases the planet's vitality, we are participating in a process that will strengthen the ecosystem and ensure the future of humankind. With my work through Ecology Action, I have strived to help people worldwide take part in this harmonious renewal through gardening. I met Cindy Conner when she attended an Ecology Action Three Day Workshop in October 2000 in Chambersburg, Pennsylvania. She was already involved in becoming food self-reliant, and she has very actively kept on equipping herself and others since then, including teaching workshops and talks for the public and at the university level.

This book is based on lots of practical experience gleaned by Cindy Conner over a 30-year period—and you can benefit from this treasure trove right now! Most important, *Grow a Sustainable Diet* is written from a fresh vantage point that places you in the action. Numerous easy-to-follow plans are given including overall garden layouts, tips on crop choices, timing plantings so your food will be ready to eat when you want to eat it, when to harvest, how long to harvest, and how to get the *most* calories and protein from the smallest area in a reasonable amount of time are shared.

The excitement and challenge of eating only what you grow one day a week, the advantage of keeping records to increase your bounty, minimizing the expense of outside inputs are explored. Even the pattern of crops to grow for the best diet and most fertile soil with the least area and effort are explored. The benefit of most insects for your garden mini-farm, plant/harvest time worksheets, garden maps with crop rotation information, and seed growing and preservation details are given.

You will be surprised, delighted and amazed about how the planning tables take the guesswork and work out of food growing each week! Food storage and preservation in a pantry, crawlspace root cellar with solar food dryers and even in a no-energy-cost cooling cabinet are described as ways to make possible eating your harvest available all year.

Everything is organized to make your learning experience easy and fun. I wish I had had this book when I began gardening and planning diets over 40 years ago. What an advantage that you have it now!

— John Jeavons
Author of *How To Grow More Vegetables*
(and Fruits, Nuts Berries, Grains and Other
Crops) Than You Ever Thought Possible
On Less Land Than You Can Imagine
January, 2014

First, a little history...

I HAVE WORKED TOWARD learning how to grow food while feeding the soil in return and I want to share what I've learned, so that others can add my experiences to their own, hopefully inspiring new ideas. Since having my first garden in 1974 at age 23, I have always been an organic gardener, avoiding chemicals and learning about soil building. I began to garden because I wanted to have a healthy family. Jarod, our oldest child, celebrated his first birthday the same summer I put in my first garden.

I became aware of the connection between what we eat and our health in a seventh grade science class. I remember seeing a chart showing nutrients, the foods that contained them, and what part of the body each nutrient (food) helped. It made quite an impression. I went on to graduate from Ohio State University with a degree in Home Economics Education. Although not my focus at the time, the courses I took in food and nutrition at Ohio State continued to influence me. My intention when I chose that major was to become an extension agent and help families be better producers at home. Little did I know then, that although that's what my calling in life would be, it wouldn't happen through the extension service. I began the adventure of a lifetime and became a stay-at-home mom. Our family filled out with four children, all the more reason to study and learn how to provide the best food possible for them.

In Hanover County, Virginia, I was often the only organic grower people knew and I would get phone calls with questions. I realized a healthy family isn't enough. I needed to work towards a healthy community and in 1992 I began selling produce, primarily lettuce, to two local restaurants. From 1993 to 1997, I was also a parent volunteer on a garden project at our children's elementary school. We began a compost operation with the food waste collected by the students in the cafeteria and leaves dropped off by parents. Garden beds were developed for each classroom teacher who wanted one. It was wonderful for the children but from that experience it was plain to me that the teachers and volunteers I was meeting had not been reading the same things I had all those years. Someone needed to be teaching the adults. Then there would be more knowledgeable people out there to work on school gardens and teach the children.

In 1998 I taught an organic vegetable gardening class through our county parks and recreation program. I highly recommend that, by the way, for those of you wanting to get started sharing your knowledge. I taught that class every winter for six years. In January 1999 I began teaching at the community college. Until that time, the horticulture department at J. Sargeant Reynolds Community College was only concerned with conventional landscaping. I taught a course in organic vegetable gardening that spring semester and taught Four Season Food Production during the fall semester. Those classes continue, although the name of the spring class is now Introduction to Biointensive Mini-Farming.

In 1997 and 1998 I added a small CSA with 10 to 14 families to my farming operation. CSA, short for community supported agriculture, is a method of marketing to families who agree at the beginning of the season to be members of the buying group for the whole season. My CSA families arrived at their appointed time each week, paying weekly for the bag of produce I had ready for each of them. In 1999 I helped start the Ashland Farmers Market in my community. In 2001, my last year to sell vegetables, I also attended the 17th St. Farmers Market in Richmond, Virginia. I had thought the way I could contribute to a healthy community was by providing nutrition-packed, chemical-free food. But I found that people needed more than that. They needed to understand what they were getting, why they should want it, how to grow it themselves, and the list went on. They needed education.

I left selling at the markets after the 2001 season hoping to be able to put more knowledgeable consumers and producers there through my teaching and researching. In the fall of 2001 I added a Growing for Market class to my teaching schedule and began work developing a Bio-intensive garden at the college. Spring semester 2002 I added Complete Diet Mini-Farming, rounding out what has become their sustainable agriculture program.

The best way to learn something is to teach it. For many years I had been studying best practices and considered myself a good organic grower, although I was never certified. After I started teaching I realized there was still more to learn. It is possible to be organic to the letter-of-the-law and still not be sustainable. One group working actively on sustainability issues was John Jeavons and his crew at Ecology Action in Willits, California. I began studying their publications and keeping records on the necessary crops. When John gave a three-day workshop in Chambersburg, Pennsylvania in October 2000, I was there. In July 2001 I traveled to Ecology Action to attend a teacher workshop and in the following year was certified as an Ecology Action GROW BIO-INTENSIVE® Sustainable Mini-Farming Teacher at the basic level. My work then became focused on what it would take to sustainably grow all of one's diet, not just food in general. In 2006 I became certified at the intermediate level of teaching. That same year I spent two weeks living in a tent at Three Sisters Farm in Pennsylvania, earning a certificate in permaculture design in a class taught by Darrell Frey.

I was learning wherever I could. Virginia Tech had begun to have field days showing new ways to manage cover crops. Actually, they were teaching an organic method of no-till. Each time you till the soil, organic matter is lost plus you contribute to hardpan, especially when using big equipment. That's why farmers embraced no-till methods. Unfortunately those methods involved killing the cover crop with herbicides, then planting. Now, instead of using herbicides, Virginia Tech and others involved in similar research were suggesting letting the cover crops grow to maturity, or almost to maturity, to get the most biomass both above and below the ground. At that point, the crop would be rolled or cut down, then planted into. Of course, they were talking farming with big tractors. Doing this work involved agricultural engineers developing roller/crimpers to roll the crop down and make sure it stayed down. In order to transplant into all that residue, new equipment had to be developed. It was important to know what crops to plant, the timing, and how to handle everything.

When I attended the first field day on this, I thought it was wonderful. Through my GROW BIOINTENSIVE work I had been growing cover crops and cutting them with a sickle for biomass for the compost pile. I was familiar with most of the crops and knew how to grow them and how much was needed. What they were showing at the field days was when to roll it or cut it to lie down as mulch and let it compost in place. It was as if all the pieces were coming together in a puzzle. I was pretty excited and I'm sure that came through when I talked to Dr. Ron Morse who headed up the project.

In 2006, I was invited to attend a series of train-the-trainer events held by Virginia Tech for cooperative extension agents, and soil and water conservation personnel. To me, finding a sustainable no-till method being promoted was huge. This was a whole new approach and I was excited to see it being taught to these people whose job it was to educate others. I wondered, however, how much of this information would get to the local farmers. In the past I had seen extension programs come and go. During the years the research was funded, there would be field days and informative meetings. Then, however, it would be written up in bulletins and buried in the files. I hoped that would not be the case with this material.

For work on a smaller scale I realized that I could translate this information into ways to manage cover crops with hand tools. Knowing the right crop to plant at the right time, and when to cut it, was all the same whether you were going through a field on a tractor or tending a garden with a sickle. Instead of roller/crimpers and no-till planting aids we just needed a sickle and a sturdy trowel. In particular, I could see how this would benefit small-scale market growers. I was going to teach this in my classes at the college, but I needed to reach even more people.

When he was in high school our youngest son, Luke, would sometimes film me in the garden and I would show the film in class. After high school Luke went to film school. Upon graduating he returned home and I realized that, with Luke's help, I could get this information to people through a video. We filmed an episode in the garden each month except August, from March through November 2007. In February 2008 we released the 66-minute DVD *Cover Crops and Compost Crops IN Your Garden*, showing it at the annual conference sponsored by the Virginia Association for Biological Farming and Virginia State University.

One thing leads to another and even before that video was finished I knew I would have to produce one about garden planning. Through my years of teaching I had developed a method of getting a whole garden plan on paper, enabling my students to all present their plans to me in the same way. This was more than just a way to make grading easier. Knowing what it takes to plan growing for the markets, I had developed a way to make that planning easier. It was information I wish I had when I was selling vegetables. At the beginning of the season I would know how many seeds and plants I would need, what was to be planted where and when, and when to expect a harvest. In January 2010 we released the video *Develop a Sustainable Vegetable Garden Plan*. It includes a 115-minute DVD, plus a CD with all the worksheets. My husband, Walt, a computer programmer, was tech support for those worksheets. In the DVD I teach garden planning for the first hour, and you meet nine other gardeners and tour their seven gardens in the second hour.

Producing those videos was intense work. Transforming the information I wanted to get across was one thing. Learning the ins and outs of video production was quite another. Luke did a great job filming and editing. Working with any other film crew would have been quite a different experience, I'm sure. To even my own surprise, once that second video was done I decided I needed to leave the college in order to address a larger community. Conveniently, our daughter Betsy was moving back to Virginia from Arkansas that year and was able to take over teaching the classes. Betsy is one of the gardeners in the garden plan video, walking you through her Arkansas market garden. Today I continue my work on sustainable diets, including how to get food all the way to the table using the least fossil fuel, and I write a blog at Home placeEarth.wordpress.com. Occasionally I get out and about to promote my work and Walt often joins me to man the Homeplace Earth booth while I'm speaking.

Our second oldest son, Travis, a talented artist and photographer, readily agreed to photograph cover crop seeds so that I could show them on the packaging of the first video. He emailed the photos to me in January, 2008. Four days later, he suddenly passed away at age 30. Since then, I have learned much about what happens when we pass from this world to the next. We are not really gone, just changed, and our energy is still here. Travis has made sure that our family and his friends have received many signs and much guidance and love from him. It has been quite a journey, these past years.

For many years people have asked me when I was going to write a book. I would always reply that I was just doing things that were already written in books and I recommended they read those books. Now, however, I'm doing things that are not in existing books and it is time to write my own. What you will read here is how I've combined the principles of GROW BIOINTENSIVE Sustainable Mini-Farming with everything else I've learned and put it into practice, in a way that works for me and which I trust can also work for you and many others. We can thank Travis for my actually getting this done. The signs and guidance from him about this have been quite strong lately. So, with help from heaven, here it is.

*All the worksheets in this book are available
for download at http://tinyurl.com/mf4a33r*

1

Sustainable Diet

THE WORD "DIET" has many connotations. Hearing that word people often think of having to eliminate certain foods from what they are eating. The definition of "diet" in my copy of *The American Heritage Dictionary of the English Language* begins "The usual food and drink of a person or animal; daily sustenance." That's what I'm talking about: getting your usual food and drink, your daily sustenance, in a way that is sustainable. The definition of "sustain" in that dictionary is "To keep in existence; maintain; prolong." A sustainable diet would be one in which the food choices are grown in a way that maintains the earth so that it can be kept in existence. Just growing this diet would replenish the earth. It's a different way of thinking about what you eat.

For the most nutrition, the least distance in space and time from the soil to the table is the best. This means that if you are following a sustainable diet you will be eating locally and seasonally. What you don't grow, you will acquire from local growers and your selection will be regional. For far too long people have had access to food shipped in from faraway places. The environmental cost of that is tremendous. Currently people have a hard time discerning what is in season in their area when the grocery store has anything they could want, all the time. Begin to notice the origins of the food in the produce section of your grocery store.

1

Growing your own or buying from local growers quickly educates you to seasonality. That doesn't mean that you will never again eat something grown elsewhere. Cultures have been trading for eons. Those items will become treats, as they once were, rather than everyday fare.

You will need to cook, a skill that is being lost with the availability of so many restaurants. Much of what is offered as prepared food does not meet the definition of a sustainable diet. When I was growing up, most people took their lunch to work in a lunchbox. Dinner together as a family was a given each night and eating in a restaurant was a rare occasion. I realize that times have changed, but if you are going to be eating food from your garden or from local growers you will be cooking it yourself. When you do go to restaurants, patronize ones that serve local food from sustainable growers. You vote for how you want the world to be with every dollar you spend, with every bite you eat.

Not everything can be eaten soon after it is picked, so you will need to learn how to preserve food. Your excess tomatoes, cucumbers, zucchini, cabbages, etc., will need some attention to save them for later months. A sustainable diet uses the least fossil fuel in this preservation. Although I still do some canning, I've turned to solar food drying and fermenting as my preferred methods of preservation. Some crops can just be stored until time to eat them. All you need to learn is the proper handling.

The world population has topped 7 billion and is rapidly increasing, stretching thin our available food growing areas. Considering that, this diet emphasizes crops that can grow the most food in the least space. The food choices in this plan are intended to meet your nutritional requirements. If you are to get all your nutrients from your garden, the one hardest to get is calories. You need to be able to fill yourself up. Potatoes, sweet potatoes, parsnips, leeks, and garlic are some of the crops that provide the most calories in the least space.

A sustainable diet feeds you, while at the same time feeding the soil and building the ecosystem. Crops are chosen for their ability to provide both food for you and food for the soil. Grains produce carbon material for compost building in the form of stalks and straw. Some crops, such as clovers and alfalfa, are grown to supply the nitrogen component to the compost pile. In a sustainable garden, over the course of the year cover crops grow in about 60 percent of the garden. Before you put this book down thinking that you can't possibly turn that much of your garden over to cover crops, remember that there are 12 months in the year and often people only use half of that time, leaving their garden to the mercy of the weeds for the other half. I'll show you how to plan those soil-building crops into your rotation.

I concentrate my efforts on staple crops and soil-building crops. The vegetable crops most people are familiar with, such as tomatoes and cucumbers, add variety to your diet and can be worked into the sustainable garden plan. Since we're talking eating only what grows best in your region, you need to pay particular attention to variety selection and seed saving. Certain varieties of each crop grow best in specific areas. If you save the seed of what does best in your niche area, you will have developed a strain of that crop that will surpass anything you can buy from a seed company.

Your meals will change with the seasons and you will become more attuned to the place you are in. Adding animal products to this diet is feasible, however the space it takes to grow the food for the animals that feed you becomes part of your nutritional footprint. My vision of the food system that develops around sustainable diets doesn't include a beef industry or a broiler industry. Instead, beef would come from the male offspring of dairy cows and from the old cows themselves. Meat from chickens would come from young roosters and old hens. We would eat less meat and prepare it in different ways. For example, rather than large pieces of fried chicken, chopped chicken and gravy would be on the menu, served over mashed potatoes or noodles. A sustainable diet can be an adventure, not a deprivation. There are so many things to try, we just have to get ourselves out of our culinary rut and do it.

I have written this book for anyone who wants to consider a sustainable diet and learn how to grow it. For those new to gardening, just growing anything is an accomplishment and you will learn more each year. Also, your soil will keep getting better and better. Many of you will be much further along this path and just need to fine-tune what you are already doing. This is a book to help you think through the whole process and decide how you can make it work for you. With this book, and others I'll suggest, you can embark on an educational journey just as if you were taking my class, except you will be moving at your own pace. I always suggest keeping a notebook with information you've gathered. Do some research on areas you need to learn more about and write up your findings for your notebook. The best way to learn is to teach others, so get some friends interested and share what you've learned with them.

What If the Trucks Stop Coming?

What if the trucks stop coming to the grocery stores? This is the question I posed to my students at the beginning of the Four Season Food Production class I taught at J. Sargeant Reynolds Community College. I wanted them to think about how they would feed themselves with homegrown or local food supplies. Their task was to complete a group project that would examine what food was available from farms within a 100-mile radius of where they lived and to estimate how much they would need for a year. Extra credit was given to each student if they brought in a highway map with circles showing distances of 25, 50, 75, and 100 miles from their home. It was essential that it be one of those fold-out maps to show all the secondary roads and small towns.

It was a real awakening for most of the students. First, they had to think about what they were eating. Some could not imagine eliminating processed food from their diet. They had to think about how much food they would need and how they would store it, with the grocery stores closed and all. In reality, those stores only have about a three or four day supply for normal times.[1] Their shelves would be empty sooner if there was a panic. Working in a group was one of the great aspects of the project. Students, who may not have interacted otherwise, were now asking each other what skills, equipment, and other resources they had. My original intent was to encourage them to get to know the local growers at the farmers markets and where their farms were located. They did that, but quickly realized that if the trucks actually did stop coming to the grocery stores, local growers wouldn't be able to meet the demand with current production. Also, not all their food needs, or desires, would be met with the current local supply. They would have to depend on each other.

This project certainly got everyone thinking and talking to one another and that is really important. If actually faced with the possibility of the trucks not coming, some people might act out of fear. They would acquire firearms and ammunition, begin to hoard food and supplies and build secret places to put their stash. I don't believe there are enough guns and ammunition to keep hungry people from helping themselves in times of peril. I have written this book to help people act out of love and compassion. Each community needs to develop resiliency to meet whatever the future brings. Many signs point to the collapse of systems as we know them now. That doesn't mean collapse of our society; it just means we have to change to meet our new circumstances. Change is part of the organic process of our life and is inevitable. You have always lived with change. The houses you live in, the clothes you wear, interests, hobbies, jobs, food choices, etc. have all evolved as you have grown into the person you are now. Certainly, we have to do everything we can individually to contribute to our own needs, but without community, we cannot survive. When things change, new opportunities open up. We can help our communities embrace the opportunities that will lead to a future with food enough for all and a healthy earth.

Making Changes

This book will help you learn how to calculate how much food you would need and how much space it would take to grow and store it. Furthermore, it will teach you to do that sustainably by building the soil and using the least fossil fuel in growing your food and getting it to the table. This is actually the food growing part of permaculture. Permaculture is a design system whereby all the energies within a system are used to maximum efficiency, the excess from one operation becoming a resource for another. Three important permaculture ethics[2] are:

1. **Care for the people:** We do what we can with our resources and increase our skills as much as we can, considering everyone who needs to be fed.

2. **Care for the earth:** Whatever we do needs to be replenishing the earth, not leaving a trail of garbage and toxic waste behind us. Everything is connected. Consider yourself living downstream from anything you have ever done or contributed to.

3. **Redistribute the surplus:** Everyone has talents in certain areas. If you think you don't, just keep an open mind and know that if you follow your heart you will discover your talents. That's where redistribute the surplus comes in. When we do what we do best, we probably have more than we need of some things and not enough of others. Share what you have through gift, trade, or sale.

Many people want to develop a small farm that will provide their family with a substantial part of their food. This book will help you understand how to do that. I know many of you are anxious to sell vegetables at the farmers markets and elsewhere. Before we can feed others, however, we need to know how to feed ourselves. I have seen farmers at the markets who have some food items for sale that their families haven't even had a chance to eat yet. Shouldn't farm families eat as well as the customers? If you take your time and grow a wide variety of food for your family first, planning out how much you need and how much you will actually harvest and when the harvest will be, and learning how to prepare it for the table in a way your family will eat, you will have undergone an educational process that you couldn't get anywhere else. Those few short years of learning will also be years of soil building and skill building. Every endeavor on a homestead seems to require additional tools and infrastructure. If you start right in as a market gardener, you will be playing catch-up for years, always needing tools, supplies, or a building that you hadn't anticipated until then. If you learn to grow for your family first, you can anticipate what it would be like to ramp up production and better plan for it. Even a small urban garden is a step in the right direction to begin your education.

Back up a bit, though. Before you can even imagine growing a large part of your food, you need to imagine eating a diet of those foods. As you travel this journey your eating habits will change, leaving behind food brought to you by the industrial food complex, and incorporating homegrown/local food eaten in season. Stay open to the possibilities and adjust your goals as you learn more. Be kind to yourself and make changes to your diet gradually.

For a few years I was on a committee at my church that partnered with a church in Haiti, providing aid. There was a yearly meeting for all the churches doing that in our diocese. One year when I attended, the lunch that was served was really hard for me to understand. Lunch consisted of sub sandwiches from a chain restaurant, offering three choices of meat or vegetarian, chips, cake and about a dozen choices of soft drinks. The dishes and silverware were disposable, in spite of there being a church kitchen available. I knew that it was volunteers who planned and gathered everything together, and since I didn't want to do that, I should have been grateful. However, I thought it would be more appropriate if the lunch consisted of something we could imagine the people we were helping in Haiti eating, or food local to us. Afterward, I wrote a letter saying so. I never received a response to that letter, but the next year lunch was provided by the local Food Not Bombs group and consisted of soup, salad, and bread. The beverages were tea, coffee, and water. It made a huge difference to me that day. I hope others realized the change that had occurred. The following year the planning committee suggested that participants bring their own non-disposable plates and cups to use. Things were definitely moving in the right direction. Sometimes, just taking another look at what you're doing, with specific goals in mind, will help you find new ways.

Right now you might think it's a good idea to grow *all* your food, and maybe you can do that. However, once you really get started you might realize that it would be better to grow some of it and support local growers for the rest. Deciding how to use the resources at your disposal efficiently is a big step. I'll give you some examples of how to use a limited growing area to your best advantage. Growing your own food is time-consuming and dirty work. You have to be ready to make a commitment to a place (your garden) and to learning new skills. I can only teach gardening in the context of the "whole system." Besides the ecosystem of what's going on in the soil and plants, it also means what is going on in your kitchen and lifestyle. If your time is filled with activities now, something will need to change to make time for gardening on a larger scale, because it is not only the gardening, but the eating that will be evolving. Cooking from the garden is different from cooking frozen or canned food from the store. Using food fresh from the garden is even an adjustment for chefs who have only ever used produce trucked in from a distance. Some people like to jump into the deep end, so to speak, and let new projects overwhelm them. Remember, however, to think of the significant others who will be on this journey with you, although not so involved. Take time to think through the changes you are making. Gradually, some things that used to seem important are not so much on your mind anymore, as your new lifestyle begins to develop.

With all that in mind, let's get started.

2

Garden Maps

MAPS HELP YOU KNOW where you are going and a map of your garden is no exception. As you are designing your map, use all space wisely. Sustainable diets strive to leave a small footprint. To assess what you have, measure the space you have to work with, draw it out on paper, and divide the garden area into beds. I like 4' wide garden beds, but some people with a shorter reach prefer 3' wide beds. The wider the bed, the more efficiently the space is used, up to a point. You will not be efficient if you are not comfortable working in a bed that is too wide.

Having wider beds gives you more growing area in proportion to your space. As you can see in Figure 2.1, in the same footprint of 700 ft², you would have 7 percent more growing space with 4' wide beds, rather than 3' wide beds. Within that same footprint, you could gain another 25 ft² by making the paths 1½' wide, rather than 2' wide.

Plan your paths. Wide paths are only needed if you are bringing in equipment or lots of people. The narrower your paths, the less you have to manage. I've found a good workable width is 18", which allows room for a rolling stool. I have trouble moving around in anything less than that. Some paths can be wider, such as when dividing sections of the garden, but not all the paths need to be wide.

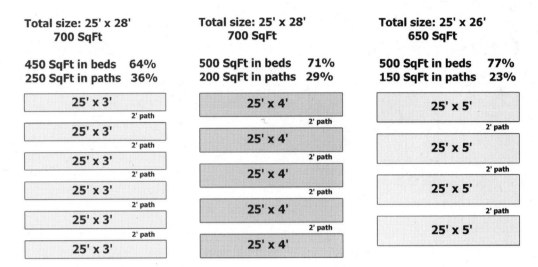

Figure 2.1. Bed Width Comparison

Have a plan for maintaining the paths from the beginning. In the narrow paths between my beds I plant Dutch white clover, or I mulch with leaves from the trees in our yard. The clover is a short-lived perennial. Occasionally it dies out and needs to be replanted. When it is growing vigorously it tends to creep into the sides of the beds. That's when it is time to harvest it for mulch or for the compost pile. I cut it with a sickle to harvest, but only have to do that a couple times over the summer. The clover keeps the weeds out, gives me a nice green carpet to be working on, and provides plenty of habitat for beneficial insects. If, instead, I put down leaves in the fall to mulch the paths, they will gradually decompose and by the next year will have become compost to toss onto the beds.

Other alternatives for mulching paths are newspaper and cardboard, but that is material you have to bring in from elsewhere. You might already have a supply as part of your circle of living. As long as they don't contain harmful chemicals to leach into your garden, these materials do a good job of keeping the weeds out of your paths. Pizza boxes are a good size to lay down in the path. Grass clippings thrown over cardboard or paper mulch make them look better, keep them from blowing away, and help them compost in place. Cardboard and newspaper are what I used

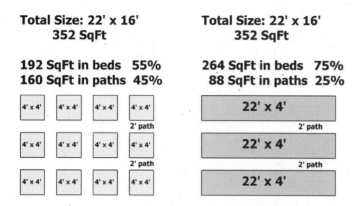

Total Size: 22' x 16'
352 SqFt

192 SqFt in beds 55%
160 SqFt in paths 45%

Total Size: 22' x 16'
352 SqFt

264 SqFt in beds 75%
88 SqFt in paths 25%

Figure 2.2. Connecting 4 × 4 Beds

before I decided to eliminate outside inputs in my paths. Recently I had an old cotton bedspread that, after forty years, had reached the end of its useful life in the house. I put that down in a path that I wouldn't be digging up anytime soon and threw some leaves on top.

If you happen to have a garden full of 4' × 4' beds, linking them together into longer beds could increase your production area by 20 percent, as shown in Figure 2.2. Besides increasing the production area, having fewer paths results in less path maintenance. Grass paths are a possibility for the wider paths, but not the narrow ones. Maneuvering a mower in tight spaces might result in damage to your plants. Grass paths require mowing and you need to be careful not to blow grass onto your beds. You don't want to be washing grass off your harvested vegetables. Bagging the grass when you mow these paths gives you mulch or compost material.

My large garden is divided into four sections of nine beds each, with 18" paths between the beds. This is the garden you see in my DVD *Cover Crops and Compost Crops IN Your Garden*. There is nothing magical about having nine beds in a row that I know of. That's just what fits into my available space and it works well for me. The sections are divided by 4' wide grass paths that I mow with a push mower, bagging the grass. When I originally planned out that garden, I imagined I would just zip through there on the riding mower, which I did in the early years. That was before the garden fence went up. As the plantings got more intense

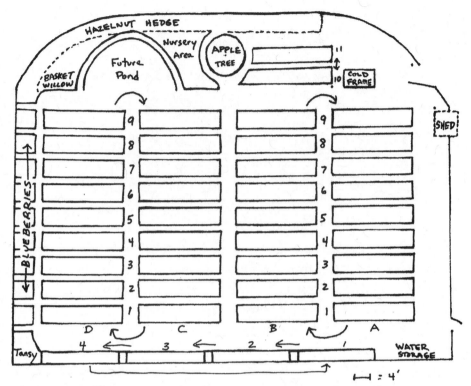

Figure 2.3. Large Garden

there was no room for the riding mower, which was okay because I couldn't bag the grass with that mower anyway. The wide paths, in addition to providing more work space and easy access between sections, are now part of my plan to supply mulch in the form of grass clippings to some of my beds.

Fall is the time at my place to redefine the bed edges, if necessary. As I harvest the last of the summer crops and plant the cover crops, I dig out the narrow paths if they've been mulched or if the white clover has died back, and toss that rich soil onto the beds. Doing this defines the path and enriches the bed. I either broadcast white clover again in the path or mulch with leaves. If you are starting a new garden, dig out the paths. That sod you remove will go to your new compost pile. Any extra soil you dig from the paths can go to build up your new beds. If it is clay that you are digging up, add it to the compost, rather than to your bed. Clay holds a lot of nutrients that can become available in the composting process.

On your garden map identify each bed with a number or letter, or both. In my garden the four main sections are designated as A, B, C, and D, plus an additional section (E) of smaller beds for perennials. The nine beds in each section are identified as A1, A2, A3, etc. Farmers make similar maps, but instead of beds they have fields, with names to identify them.

If you anticipate predators you may want to plan for a fence. I'll address fences in Chapter 12. My large garden has a good fence now, but it had none when I started out. Even without a fence, you can establish a border. A border defines your space and could be an area surrounding your garden, planted to annual or perennial plants that will enhance your ecosystem, attracting beneficial insects that keep the harmful ones in balance. More about that in Chapter 6. A border could be allowed to develop into more of a hedge, planted to one or more types of bushes or small trees. A hedge is something to keep in mind for the future as you plan your space. The height of a hedge needs to be considered so it doesn't shade your vegetable beds. As you can see on the map (Figure 2.3), I have a hazelnut hedge planted on the north side of my garden.

I run my garden beds east to west. In my large garden, each bed is 4' × 20'. In 1985 I began with 12 beds, each 5' × 20' and planted in pairs. Each pair of beds had a 2' mulched path between the beds and was surrounded by a 4' grass path. The idea, which I had read about in a magazine article, was to allow space for a garden cart along each bed for bringing in mulch. It seemed like a good idea at the time, but it didn't take long for me to realize that I only needed 18" between the beds and that if I wanted to bring leaves in for mulch, it could be easily managed from the short sides. I also discovered I'm more comfortable working in a 4' wide bed than a 5' one. I took out those wide grass paths, making room for more beds, and made the paths between the beds 18". The new plan increased my growing area within the same space and decreased the area needed to be maintained as paths.

Wire grass, a type of Bermuda grass, is a problem in my area anywhere there is grass, and my garden is no exception. It shows up whether we want it to or not, spreading by both rhizomes and stolons. Limiting the grass along the bed edges limits the threat of wire grass creeping in. Often, people want to surround their beds with wood boxes. That's not something I recommend, unless you have a really good reason. Doing it because that's how you think you start a garden is not a good reason. Doing it to exclude wire grass is an even worse reason. If you have wire grass, a box around your beds is not going to keep it out. In fact, it prolongs the problem because parts of the plant will sit right under the edge of the box to keep coming back. The same goes for other barriers you might spend time and money on. Having narrow, well maintained paths leaves no place for the wire grass to grow.

Don't hesitate to make changes if you find things don't work as well as you originally planned. A bed length of 20' works well for me and fits my space nicely. When I was selling vegetables, I had an additional garden with beds 4' × 75'. A market grower friend of mine had beds 4' × 200'.

Plan Outside the Box

You don't have to stick with rectangles for your garden beds. My smaller garden began with rows of 4' wide beds in 1984. The next year I redesigned it, imagining it would be filled with flowers and herbs and look nice when we pulled in the driveway. I had started the larger garden for my main vegetable production. The new design for the small garden, which I found in a book, was not practical for me. That redesign might have been nice if I had more time to work with it, but our fourth child was born the following year. With a busy family life, I was doing well just to keep up with vegetable production in the larger garden. Just because it is in a book, doesn't mean it will work for you. That goes for this book also. What I want to do here is to teach you to think. Yes, by all means, try my ideas and see if they work for you. If not, maybe just a little tweaking is all you need, but you need a starting point: a frame of reference.

A few more years went by and eventually we needed to have more driveway space to allow for parking and turning around. The road we live on had become too busy to back out into safely. I enrolled in a landscape design class being taught at the local high school to help with

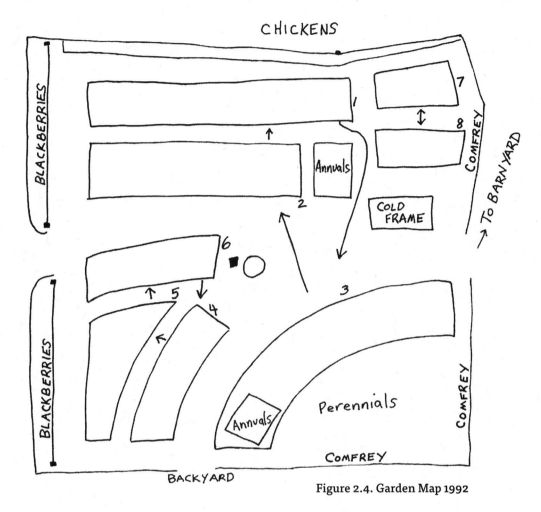

Figure 2.4. Garden Map 1992

this planning. Mapping out the area on graph paper helped me decide that the place to add the new parking area was beside the small garden and would interfere with the garden path that went through the middle. When I moved the path over a few feet, everything changed. I put in a curved path that took me from the backyard to the barnyard. From there, other paths fell into place, for a time (see Figure 2.4). Every few years I find some reason or other to make changes in that garden, but that backyard-to-barnyard path stays put. If you have a path in your yard, put a garden around it. Decide where you would naturally be walking with a purpose and plan your beds around that.

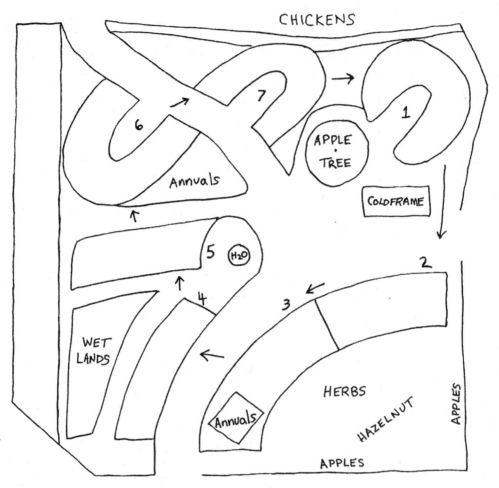

CHICKENS

6

7

1

APPLE
• TREE

COLDFRAME

Annuals

5 (H2O)

2

4

3

WET
LANDS

Annuals

HERBS

HAZELNUT

APPLES

APPLES

Figure 2.5. Garden Map 2009

That garden has always had some areas for flowers and herbs, in addition to the vegetable beds. The redesign had three 100 ft² beds for vegetables that rotated together and three 50 ft² vegetable beds that rotated, along with a couple of smaller spots for vegetables and a coldframe. Along the north side is the chicken pen fence. When I divided the chicken pen into different areas, one for composting, I decided to make a gate from the garden into the compost area of the pen. For easier access, I needed a new path. This new design gave me an opportunity to put in some keyhole beds (Figure 2.5). It had seven 50 ft² beds in

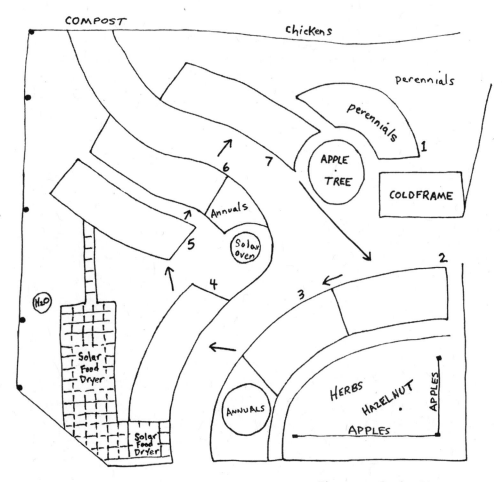

Figure 2.6. Garden Map 2013

rotation, including one for compost. I added apples to cordon in the
southeast corner. As they grew they were trained along horizontal sup-
ports to form a fence.

A few years later I needed space for the solar dryers I'd built so we
wouldn't have to move them to mow. In addition, although the keyhole
beds were okay, the 4' wide beds were easier for me to manage. This
called for another redesign. I've now made a space for the solar dryers
in this garden and converted the two keyholes beside the path to the
chicken composting area to 4' wide beds. The keyhole bed on the other

side of the garden is also gone. I no longer devote a bed for compost in this garden. With access to the chicken composting area right at hand, that's where the compost materials go. The 3' × 6' coldframe needed to be rebuilt. The new one is 4' × 8'. Bed #1 is smaller now and out of the rotation. Strawberries are planned for that spot. That upper right corner is still evolving—food forest style. A grape vine is going in, for sure, and some other perennials.

Only you can decide what size to make your garden beds. A uniform area in each bed, no matter what shape, makes it easier to plan rotations and to anticipate seed needs and yields, which I'll talk about in coming chapters.

Permaculture Plan

Now, I want you to draw another map. This time make it a map of your whole property. Your house, driveway, and outbuildings will be in it. *You* will be in it. You will be considering all of it when you grow your sustainable diet. I call this map your permaculture plan. You will remember that in Chapter 1 I defined permaculture as a design system whereby all the energies within a system are used to maximum efficiency. On this map you can identify niches for certain things that are just the right size and microclimate, even if they aren't in your main garden area. A protected area for a fig tree or a spot for herbs by the back door are examples of this. You can also use this map to help identify places to store equipment and crops. The *Earth User's Guide to Permaculture* is a good resource to consult when making a map such as this.

We usually look out at our property from our houses. In order to really look at your property with new eyes, go to different corners of the property and look from there. Take plenty of pictures. Go over to your neighbor's house and look from there. Your home and gardens will definitely look different from their yard. You might also discover that the junk you put behind your shed so you wouldn't see it is in full view of the neighbors when they sit on their deck. When you are cleaning out your gutters, take a camera up that ladder with you and take pictures of your property from above. We have a sandbox that we maintain for young visitors and it has a roof over top for children to climb up to. Atop that sandbox is my favorite place to photograph my garden at various times of the growing season.

Observe your property at different times of the year. If you live in an area that gets snow, take pictures as the snow melts. Where does it melt first? That's your warm spot. Where does it linger after everywhere else looks like spring is coming? That is not the place to plant your early spring crops. As the sun gets higher in the sky, trees and buildings will cast shorter shadows. I have my solar dryers in a great place in the garden, but by September the big maple tree in the backyard begins to cast a longer shadow and I have to move the dryers to catch a full day of sun. On the other hand, if you thought all you had was shade in your yard because of the trees, look again once the leaves have dropped. You might be able to plant a winter garden in the sun, having greens for your table during the cold months. By the way, make sure the trees are on your permaculture plan.

When I first planned my large garden, compost bins made of pallets lined the north side. It made sense because it was easy access when I brought in compost materials from elsewhere, primarily leaves and animal bedding. Once I got serious about sustainability, I realized I wanted to limit outside inputs and began growing my own compost materials, which meant adding grains to my garden. Everything was now being moved around within the garden, rather than being hauled in. Before I knew it, I didn't have to bring the garden cart in anymore. As the biomass from the beds was harvested, it was carried to the compost piles and finished compost was delivered to the beds in buckets.

After seeing how well my butternut squash grew over my compost pile, I began to rethink how I managed compost. There was obviously fertility leaching away under those compost piles on the north side of the garden and I wanted to harvest it to feed my crops. I got rid of the bins and planned my compost to become part of the garden rotations. The compost piles are now on garden beds, with butternut squash growing over the pile that was built the previous fall. That pile matures over the summer, underneath the squash plants, and is ready to spread in September and October. Any compost that needs to be held longer is turned to the next bed in the fall and the new piles are built on that new compost bed. With the compost bins gone, I bumped out the fence on the north side and have a hazelnut (filbert) hedge there. I also found room on that north side for a garden washing station, two more garden beds, an apple tree, basket willow, and a spot I've reserved to dig a small pond when I find the time (see Figure 2.3). With my compost, just as with the lunch at the Haiti gathering that I told you about in Chapter 1, it only took looking at the project in a new way. Stay open to the possibilities and new ones will present themselves, right before your eyes. All you have to do is recognize them as such. The more you make yourself part of the plan, the easier it will be to identify these things.

Tools for Map Making

You can start with graph paper, a ruler, pencils, and an eraser. You're going to need that eraser. Your maps will reflect what you put into them. If you are mapping out your beds, you will definitely need to measure carefully and be accurate. If you are making a map of your property, you decide how exact you need to be. If you know the length of your stride, pacing off your boundaries might be enough. I remember in my high school marching band we had to all take the same size step. Whatever yours is will be your measure. If you do a really good job on this map, it will be more help to you later.

I've found a 100' tape measure to be a wonderful tool to have. It has a fiberglass tape with inches on one side and decimal designations on the other. The end has a hole that I put a screwdriver through to anchor it in the ground when I don't have anyone on the other end to hold it. Although in a perfect world it would be clean and dry always, it survives nicely if it is occasionally wound up wet or dirty, since the tape is not enclosed. You can find one at building supply stores. A ruler is good, but an engineer's scale is even better, particularly if you are drawing on plain paper, rather than graph paper. If you really like this sort of thing you could sign up for a landscape drawing class somewhere, but graph paper, pencil, eraser, and a ruler will get you started.

It is helpful to choose a reference spot when making your maps. It might be your house or a fence that's already there. It may even be just a post. I've found it extremely helpful to have a metal fencepost at the corners of each section in my large garden. If I need to realign the beds I can quickly run a string and know where the corners are. If you don't already have a fence or post in your garden, now is the time to plan it in. Besides being a point to measure from, posts and fences give birds landing spots to check out the area for insects to eat.

Make sure to show the scale—how many feet in one inch—on the map. To get the detail you want for your permaculture plan, you may have to tape pieces of graph paper together. If your map is bigger than what you can copy on a copy machine, fold it and copy the pieces. You can tape those pieces together later, maybe on a backing of poster board. If you are copying your map in pieces, while you are making copies of the original plan at 100 percent to have an exact copy, make some smaller ones. Figure out what size you need so the resulting pieces will all fit on a standard size sheet of paper. Cut out what you need from the smaller copies, tape them on their new size paper, and make copies of that. You can make many sizes to play with. The more familiar you are with the copy machines at your office supply store, the better. You can always ask the employees to make the different size copies for you, but it is good to learn to do it yourself.

An 8½ × 11 size is nice to put in your notebook. You might want a larger size with color added (I use colored pencils) to put in a poster sized frame to hang on the wall. If you don't have a map that shows the whole of your property, you will always be leaving something out in your mind. It also helps to see things in proper proportion to everything else on your property. You could cover copies of your garden map and

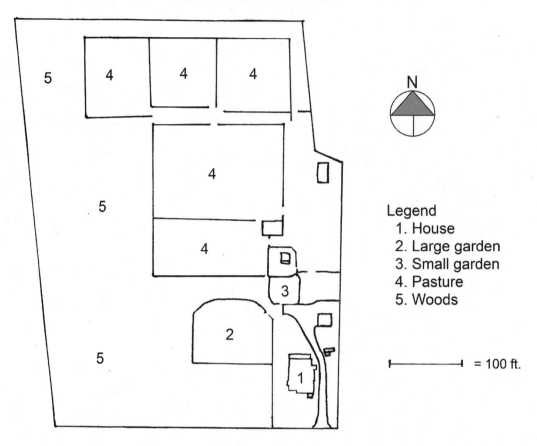

Figure 2.7. Base Permaculture Map for Sunfield Farm

permaculture plan with clear contact paper. Using wipe-off markers, you can have fun thinking up new plans. When you come up with something you like, draw it on one of those extra copies you made. Your family members might even want to get involved with that.

Leave a wild spot on your property when you are planning. That would be an area that is left alone and not cultivated. If you are living where "wild" might be frowned upon, some bushes or a tree with bushes will do. Anywhere that is not messed with too often will provide habitat for beneficial insects. Those bushes could be something that flower early, providing nectar for the bees. (Having a perennial flower bed could also fill this niche.) If your neighbors live close and have phobias about insects, maybe it might be best not to mention attracting bees and bugs.

Wild areas help to maintain diversity. If they connect from one property to another they become wildlife corridors—highways for the wildlife. Not cultivating from property line to property line is a start. Some communities plan for undeveloped space. Unfortunately, sometimes residents do not recognize the value of those areas.

The base permaculture map of our place, Sunfield Farm, is shown above (Figure 2.7). This is the starting point of all the plans and only shows the property lines, fences, and buildings. At times I've made larger maps of portions of this map to show more detail when planning projects. I've done that for each of the gardens, the backyard, the back pasture area, and the barnyard. If you own your house you may have a copy of the plot plan that resulted from the survey that was done when you moved in. If you are not skilled at map making you could just use that as your base map.

3

Crop Choices

IF YOU ALREADY HAVE A GARDEN, you already have a list of
favorite crops you want to grow. In developing your plan
for a sustainable diet, you will want to consider crops that will grow
the most food in the least space. John Jeavons and the folks at Ecology
Action in California have done much research in that area and have
documented their work in Jeavon's book *How to Grow More Vegetables.*[1]
That book will be a helpful resource, in addition to this book, for this
journey. Their gardening method is called GROW BIOINTENSIVE®.
When you see it written in all capital letters like that, you know that all
eight elements of the method are being followed. Those elements are
deep soil preparation, use of compost, intensive planting, companion
planting, carbon crops, calorie crops, open-pollinated seeds, and the
whole system. I have been studying GROW BIOINTENSIVE for some
time and think of all those things as best management practices for any
garden.

According to Jeavons, if you were to eat only what you grew in your
garden, you would need to give serious attention to getting enough calories, protein, and calcium. Most likely, your diet would include other
local foods; occasionally some not-so-local foods; and some animal

products, which we'll talk about later. But for now, let's suppose that everything in your diet is from your garden.

Growing Calories

With calories being the biggest limiting factor if all your nutrient needs were to come from your garden, we'll take a look at that first. The foods on Jeavons' list of crops producing the most calories in the least space are garlic, Jerusalem artichokes, leeks, parsnips, potatoes, salsify, and sweet potatoes. In my garden, potatoes, sweet potatoes, and garlic are important crops. I have not taken the time to learn the nuances of leeks, parsnips, and salsify. Jerusalem artichokes were given to me many years ago by my friend Chester. It is one of his favorite foods. Although they haven't become a staple at our table, I cook them occasionally and use them in ferment. They are great to dig through the winter, and with Chester's encouragement, continue to be a part of my garden. I planted them at home and in the college garden where I was teaching. As a result, my students took some home to plant. When the deer ate Chester's plants I was able to give some back to him. Being generous with what we have always pays dividends in the long run. An aspect of a sustainable diet is making sure of our supply. We can do that by sharing seeds and plants so more people are growing them.

Potatoes will give you the most calories per square foot planted of anything you will grow. Eating a diet of only potatoes could be toxic due to an excess of potassium. On the other hand, if you need potassium, eat more potatoes. Having fermented foods, such as sauerkraut, in your diet could help with detoxification.[2] In *Nutrition and Physical Degeneration*, Weston A. Price wrote about the Quechua Indians of Peru who ate mostly potatoes dipped in a slurry of kaolin clay.[3] The practice of eating earthy substances, such as clay, is called *geophagy*. Eating certain kinds of clay with wild potatoes was a practice of the Indians in the American Southwest and in Mexico, as well as those living in the Central Andean Altiplano.[4] The clay would have helped rid the body of toxins. Of course, today's varieties are not necessarily the same as those early wild potatoes. Studying indigenous diets is interesting if you want to grow all your own food, but it is important to recognize all aspects of what those earlier populations were eating and how they were eating it. Our culture has lost some of the practices that were important in bringing food to the table. Sometimes they are the keys that we need to be successful in our endeavors. I'm not advocating eating clay with your potatoes, but offering an understanding of how a culture might have subsisted on a large quantity of the varieties that were available to them.

Sweet potatoes are one of the most nutritious foods you can grow.[5] In our area we plant Irish potatoes in the spring and harvest them in the summer. Sweet potatoes grow in the hot months and are harvested in the fall, just before frost. They are even easier to keep than Irish potatoes, since exposure to light doesn't turn them green. They can usually sit in a box or basket in the corner of your house all through the winter until you need them. Sweet potatoes may produce a little less per 100 ft² than potatoes, but have a few more calories per pound.

Garlic is a good calorie producer in a small space. We don't eat it in the quantities that we do potatoes, and for good reason, but it only takes a little garlic regularly in your diet to keep you healthy. There was a woman who visited the farmers market when I sold there. She was in her nineties at the time and told me she took no medications and she ate garlic daily. She also credited being active in her church to her secret of longevity. Over one hundred years old now, she is still kicking around and I see her picture in the paper now and again, celebrating her age and spirit.

As for leeks, parsnips, and salsify, you'll have to look to others for guidance on those crops. One of those folks would be John Seymour, author of *The New Self-Sufficient Gardener*. Although he is no longer on this earth, his writings continue to guide many people along this path. Parsnips, leeks, and salsify are much bigger crops in the UK, where Seymour is from, than here in Virginia. At the end of the winter, if you still have parsnips in the ground, dig them up and make wine. Directions are in his book.

Even if you aren't growing all your own food, being able to a make a filling meal occasionally, with only ingredients from your garden, is satisfying in many ways. Jeavons lists peanuts, soybeans, dry beans, cassava, and burdock as crops that can give you a lot of calories in a small amount of food eaten, but those calories take more space to produce than potatoes. I know that cassava is a staple crop in tropical areas,[6] but I know nothing about burdock. If you are going for a sustainable diet, where you are planted on this earth will determine what you are growing and eating. I'll address the legumes (peanuts, soybeans, dry beans) for their protein importance.

Growing Protein

All vegetables contain some protein. If you concentrate on growing and eating the calorie crops I just mentioned, you will be getting a lot of protein. You will also be eating a lot of food per day. Beans and peanuts, good to combine with grains in your diet, have a lot of calories in a small amount of food but at the cost of space in the garden. Grains also produce lots of calories per pound of food but not so many per square foot in the garden, although they have the big advantage of producing biomass for compost making and mulching. When it comes to protein, beans do not contain all the amino acids we need, but they don't have to. Nature has conveniently provided the amino acids that beans are lacking in grains, and vice versa. Including both beans and grains in your diet gives you what you need. One needs only to look at traditional diets to find examples of this—tortillas and beans, beans and rice, cornbread and beans. Even good old peanut butter on whole wheat bread is an example. Animal products have all the amino acids. Eating only beans or only grains would leave you lacking. Eating both beans and grains, particularly with the addition of even a small amount of milk, meat, or cheese, assures you are meeting your body's requirements for protein. Eating a varied diet allows you to gather those amino acids from many sources.

Favas are highly recommended at Ecology Action[7] both as a diet crop (beans to eat) and as a source of green biomass for compost making. Be aware that some people are allergic to fava beans. When I tried to grow favas at my place in Zone 7,[8] the blooms would fall off prematurely during spring hot spells and didn't yield many beans. I tried pinto beans in my garden, but couldn't get much of a yield of dry beans. Pintos are popular in the gardens at Ecology Action. The climate is different there in Willits, California, with hot days, cool nights, and little humidity. Here in Virginia we have hot days, hot nights, and high humidity. I began growing cowpeas after a number of dry years. Traditionally, cowpeas, also known as Southern peas, are a good crop for my region. According to the Master Charts in *How to Grow More Vegetables*, cowpeas produce less than many other varieties of beans. The charts are a good place to go for information and guidance, but ultimately, you have to try different things and find out what does best in your area. It turns out that cowpeas are in a different bean family than other dry beans and the bean beetles that bother those other beans aren't interested in cowpeas. Furthermore, my plants yielded from 12 to 20 seeds per pod! Cowpeas have become my dry bean of choice to grow for food for our table. In my own chart of yields, cowpeas outshine pinto beans, by far.

To make beans more area-efficient (more calories produced in less space) you could interplant them with corn, growing pole varieties of beans up the cornstalks. Remember the stories of the Three Sisters? Corn, beans, and squash were key staples for the American Indians. That brings us to grains in your garden. Corn is easy. You've probably already grown sweet corn. If you are interested in getting the maximum food value from your garden you will want to grow corn all the way out to dry seed, so you would choose varieties best for grinding to cornmeal. If you aren't ready to be grinding cornmeal, but you want to experience growing corn out to dry seed, try popcorn. Popcorn is a good way to get acquainted with harvesting dried corn and cornstalks.

Rye and wheat are great for covering your garden through the winter. Cereal rye (not rye grass) and winter wheat are planted in the fall and harvested the following summer, in time for a crop of something else to follow. In my area in Virginia in Zone 7, I harvest these grains in mid-June. Spring wheat is planted in areas with winters too cold for fall-planted wheat. These grains are good additions to your diet and provide biomass in the form of straw for compost making. You could grow rye and wheat and learn to make your own sourdough bread. Corn, wheat, and rye are also choices as compost crops for their stalks and straw, providing food for us and food for the soil, but more about that in Chapter 5.

You may have read about the wonders of amaranth. I have and it sounds like a terrific crop. I even grew it to see what it was like. It has very small seeds and if you aren't careful, it can become a weed in your garden. In fact, amaranth is related to pigweed. I can't grow everything and decided that corn, rye, and wheat would be the grains in my garden.

I have peanuts in my garden as a soil builder, in addition to being a diet crop. We eat peanuts from the shell, often roasting them first. Making peanut butter from them is too much bother for me. The dried foliage from the peanuts—peanut hay—can be used to feed your compost pile or small livestock, such as goats or rabbits.

In my discussion here, I should mention soybeans. I avoid soy in my diet because it is high in phytates and contains enzyme inhibitors, which can lead to protein assimilation problems. Unless soy is consumed as a fermented product, such as miso, natto, or tempeh, it can be damaging to your health if you consume it regularly.[9] You can certainly grow soybeans in your garden as a soil builder. In that case, it would be grown as a green biomass crop for compost material, harvested when it is flowering. The conventional food industry in this country has convinced a huge population that soy is necessary to their existence. The Weston A. Price Foundation is the place to find more information about that issue. If you grow soy out to eat, learn to ferment it. Much of the soy used in the food industry is disguised to make it taste like something else. Learn to celebrate the unique flavors of each food. Food should not have to be made to taste like something else.

Growing Calcium

Most likely, dairy products are the first things that come to mind when thinking of calcium. If a cow or goat is not in your plan, you can put calcium on your plate in the form of leafy greens. Collards are loaded with calcium, with a cup of cooked collards containing about as much calcium as a cup of cow's or goat's milk. Kale is also high in calcium. Parsley has as much calcium as collards, however it is also high in oxalic acid. Oxalic acid will tie up the calcium in your diet, but it can be neutralized by cooking. For that reason, you should avoid eating raw large quantities of foods that are high in oxalic acid. The spinach family crops (spinach, beet greens, and Swiss chard) are high in oxalic acid.

Growing and eating foods high in calcium is only part of the story. You have probably heard that it is important to have enough vitamin D to work with the calcium and you can get vitamin D from being in the sun. D is a fat soluble vitamin (as are A, E, and K) so you need fat as a catalyst to help things along. A bit of oil or butter on your greens and other vegetables helps with assimilating all of these fat soluble vitamins. Peanuts and hazelnuts (filberts) are sources of both calcium and fat. Hazelnuts grow on trees and can become part of your permaculture plan as you expand beyond the vegetable garden.

Oils and Sweeteners

Peanuts and hazelnuts that provide calories, calcium, and fat, as well as protein, can be pressed for oil. Sunflower and pumpkin seeds can also be pressed for oil, but you need to make sure you are growing the right varieties for that. Growing crops just for cooking oil takes a lot of space. A sustainable diet would use limited amounts of oil for that reason. Other nut trees, such as pecan, English walnut, and black walnut might be part of your permaculture homestead. It takes time to grow them, and you have to put in some work to shell them, but the nuts can be pressed for oil.

Honey is a good product on a permaculture homestead. Once you become aware of the nectar sources for your bees, you will want to keep something blooming in your diverse plantings. Buckwheat is good as a short-term filler in your summer garden beds and the bees love it. Most of your cover crops become bee food. Keeping the bees in mind as you plan borders and companion plants will also help. Trees and shrubs provide an early nectar flow for the bees. This is outside of your annual garden beds but within your permaculture plan.

Maple syrup and sorghum syrup can be produced on your land. You have to take into account the fuel that is needed to cook the sap down to syrup to determine the ecological footprint of these products. I grew up in maple syrup country in northeast Ohio. I don't know how many small sugar bushes are still operating, but we buy local maple syrup when we visit. Of course, you need sugar maple trees for that. Sorghum can be grown in your garden for both grain and syrup. I have had little success with the syrup part, but grow it for the grain. Considering the ecological footprint of syrups, just like cooking oil, you would be limiting the amount you use in a sustainable diet.

Other Crops

Certainly, we are going to be targeting crops that can give us the most of what we need in as small a space as we can. However, as with everything else, we have to look at the whole picture. Learn as much as you can about diet and nutrition. Sometimes it is the small addition to a dish that makes it special to the eye and to the appetite and is the key to unlocking all the nutrients. My garden wouldn't be without onions. They have many health benefits and add much flavor to dishes. Tomatoes and peppers are also high on my list to include. I've mentioned collards and kale for greens to grow, but all the cabbage family is important.

Winter squash is also on my list of important crops to grow. It can go in the Three Sisters bed with the corn and beans or it can wander in your pathways. Winter squash is loaded with nutrition, and best of all, it can just keep all by itself until you want to eat it.

More Planning Tools

The Resilient Gardener by Carol Deppe is an excellent book to read on the subject of choosing crops to sustain you. In this book she talks in depth about the need to plan a garden to grow staple crops—in particular, potatoes, corn, beans, squash, and eggs. Deppe is a plant breeder and has developed seed varieties to suit her diet and her climate.

Ecology Action has some publications, available through Bountiful Gardens, that address planning a sustainable diet. Their Self-Teaching Mini-Series #36 *An Experimental 33-Bed GROW BIOINTENSIVE® Mini-Farm: Growing Complete Fertility, Nutrition and Income* has some good information on diet crops if your food is to be limited to what is grown in your garden.

If you really like to plan things by the numbers, take a look at the Ecology Action publication #31 *Designing a GROW BIOINTENSIVE® Sustainable Mini-Farm*. It has forms that are used by the Intermediate-level Ecology Action certified teachers when submitting their garden plans each year. It helps you plan crops that provide the necessary nutrients to a diet and enough cover crops to feed back to your soil. Not everyone is ready for that level of planning. Just because you plan a diet on paper doesn't mean it is a diet that is practical to eat every day. Use these things as planning tools to help you think through the issues and develop a diet and farm plan that addresses the needs of your family.

Varieties Specific to Your Region and Your Needs

Maybe you aren't growing all of your food and will be obtaining your staple crops from local suppliers who are growing them using ecologically sound practices. No matter what you grow, you will learn more each year. In Chapter 5, I will discuss cover crops and how to use them in your plan to increase soil fertility. For the vegetable crops you choose, do some homework to discover which varieties will have a better chance of doing well in your area. The varieties of squash that do well for Carol Deppe in Oregon, or Cindy Conner in Virginia, may or may not be the ones that do well for you. That's not to say you can't be adventurous. I live in Virginia, an hour from Thomas Jefferson's Monticello. Jefferson's adventuresome spirit with plants is well known. He tried many things and not all would be called successful. Remember, however, there are no mistakes—only learning experiences.

I already told you of my experience with pinto beans versus cowpeas. Do as much reading as you can about what is local to your area. If you can, find a seed company that caters to your region. Ira Wallace of Southern Exposure Seed Exchange is the author of *The Timber Press Guide to Vegetable Gardening in the Southeast* which will help those in the southeast region of the United States find the best crops to grow in their area and determine when and how to plant them. Watch for similar books written about your region.

Talk with other growers. My daughter had good luck with Turkey Craw beans. She raved about them and gave me some seeds. I just might give them a try. One year I tried Seminole squash because my friend Mark talked it up. I didn't like it as well as he did and went back to my tried and true butternut.

Make sure you know what to expect in the varieties you select. If you are growing beans—how tall are they? If they are bush varieties, you just have to plant. If they are space-saving pole varieties, you need to have something for them to climb on. It could be the corn stalks, and in that case look for bean varieties that enjoy a little shade among the cornstalks. I've seen some beans designated as half-runners. You're on your own to figure out what that means.

You'll find melon and tomato varieties suitable for areas with a shorter hot season and other varieties that need every bit of heat that summer can provide. How hot (or cool) are your days? Did you know that a consistent minimum of 60°F can have an effect on your plants? From my temperature records I can tell that our nighttime lows are consistently 60°F or above from the first week in June until the first week in September. If your summer nights are consistently *below* 60°F, as they are in Willits, California, different varieties of the same crops would be thriving. You can begin now to keep temperature and precipitation records. It is so easy to forget just how wet, dry, hot, or cold it has been, even just since last month. Having a record to look back on can help you understand what is going on with your crops. Having a record for many years is even more helpful. I have developed two forms to help with that (Figures 3.1 and 3.2 on the following pages).

Keep in mind that the more you know, the more you don't know. One year I decided that I would put six years of precipitation records on a graph. Consistently, without a doubt, October was the driest month of the year. Until that year, that is. As soon as I established that "fact" we had the wettest October I have experienced! Keeping weather records helps you become more in-tune with your place, in spite of occasional fluctuations.

TEMPERATURES

_____ (year)

Figure 3.1. Temperatures

	Jan	Feb	Mar	Apr	May	June	July	Aug	Sept	Oct	Nov	Dec	
1													1
2													2
3													3
4													4
5													5
6													6
7													7
8													8
9													9
10													10
11													11
12													12
13													13
14													14
15													15
16													16
17													17
18													18
19													19
20													20
21													21
22													22
23													23
24													24
25													25
26													26
27													27
28													28
29													29
30													30
31													31

Homeplace Earth
Education and Design for a Sustainable World
www.HomeplaceEarth.com

Download this worksheet at http://tinyurl.com/mf4a33r

PRECIPITATION

_____ (year) Total for the year [＿＿＿＿]

	Jan	Feb	Mar	Apr	May	June	July	Aug	Sept	Oct	Nov	Dec	
1													1
2													2
3													3
4													4
5													5
6													6
7													7
8													8
9													9
10													10
11													11
12													12
13													13
14													14
15													15
16													16
17													17
18													18
19													19
20													20
21													21
22													22
23													23
24													24
25													25
26													26
27													27
28													28
29													29
30													30
31													31

Monthly Totals												

Figure 3.2. Precipitation

Download this worksheet at http://tinyurl.com/mf4a33r

4

How Much to Grow

You've determined how much space you have and made your garden map. Also, you know what crops you want to put there. The next thing to consider is how much of everything you want to grow. Since we are talking about working through a diet, think of how much of each thing you need for your diet plan. Once you know how many pounds of each crop your family needs for a year, you can work from there. If you are wondering how much people usually eat of each crop for the year, there is a column for that called *Pounds Consumed per Year by Average Person in U.S.* in the Master Charts found in *How to Grow More Vegetables.*

Make a list of your crops, the number of pounds you need for the year, your target yield, and the space it will take to grow that amount. Include columns for the area you have allotted for that crop and the expected yield from that area. Have a column for your actual yield per 100 ft². That figure can only be filled in from your experience. If you can't grow the whole amount that you project that you need, having all the numbers on a worksheet together will help for future planning. You can use the worksheet that I've designed (Figure 4.1) or design your own.

The Master Charts mentioned above are a good resource for the yield estimates. There is a column in the Master Charts that shows the *Possible*

39

GROW BIOINTENSIVE Yield in Pounds per 100 Square-Foot Planting for each crop. Three numbers are given: designating what might be expected at the beginning level, an intermediate yield that might be reached after much soil building has been accomplished, and the high yield. Most people could come close to the beginning yield. Only very few will reach the high yield. Another column in the Master Charts shows the *Average U.S. Yield in Pounds per 100 Square Feet*. The yield you will achieve will depend on many factors including your soil, climate, and management skills. Using these figures as a guide, decide on a realistic yield to use as a target in your plan. Of course, if you already know how much to expect

Figure 4.1.
How Much to Grow

Crop	A lbs per week	B # of weeks	C total lbs needed AxB	D target yield / 100 ft²	E ft² needed (C÷D)x100	F actual garden area	G estimated total yield	H actual yield / 100 ft² (DxF)÷100	I Comments	J calories per pound	K protein per pound	L calcium per pound

find more nutrition tables for download at USDA Nutrient Database for Standard reference   ndb.nal.usda.gov/ndb/search/list   (there are 453.6 grams/lb)

Homeplace Earth
Education and Design for a Sustainable World
www.HomeplaceEarth.com

Download this worksheet at http://tinyurl.com/mf4a33r

for some of your crops, you can use your own yield numbers. Eventually, with diligent record keeping, you will have your own yield figures for all your crops. For fun you could do some calculations to see how much space you could save if your target yield was higher. As the quality of your soil increases, along with increases in your skills and understanding, most likely your yields will increase—to a point, of course.

A sustainable diet is a seasonal diet. You won't necessarily be planning to have each food on your table for every week in the year. Some things are best eaten fresh from the garden. On the other hand, you will want to have enough staple crops for every week. If you have never stopped to think about how much your family needs each week, now is the time to do that. My worksheet has columns for how many pounds are needed of each crop per week and how many weeks you would be eating it during the year. You don't have a suitable scale at home to figure this out? That's okay because you can use the one at the grocery store. In the produce section, pick out how much you would need to cook up for a meal and put it on the scale. Multiply that by how many times you would serve that much during the week. Are you eating it only when it is in season? How many weeks would that be? Multiply the number of weeks by how much food is needed per week and you will know how much is needed for the year. Having a good scale in your kitchen is helpful for many things. Working by count or by bunches has its limits.

This all assumes that what you are eating is food that comes from the garden. If what you are eating currently comes from boxes and cans, you will have a transition period. As I mentioned in Chapter 1, it is an adjustment to handle food in its raw state in your kitchen. There are no instructions on the packaging. In fact, there is no packaging. It opens the opportunity for creativity.

There is always something to remember about a previous harvest, favorite (or not so favorite) varieties, etc. Record those things in the space for comments. On the worksheet I've included columns for the amount of calories, protein, and calcium per pound of each crop. These figures are in the Master Charts where you found the yields. If you are taking those nutrients into account as you plan your sustainable diet, it is handy to have those figures available for reference and comparison.

If not, you might use those columns for the things you are comparing with your crops. Some information on the worksheet will be carried over from year to year. Write that in once and make copies of the worksheets. That will save you from having to add it each time.

Biosphere 2

We can make our plans and work toward their fulfillment. Meanwhile, if our community food systems are also being developed, we have something to fall back on if we need it. But what if we had to live off only what we grew? That's what was planned for the eight person crew of Biosphere 2, a self-contained sealed environment experiment that took place in Arizona. It was thought that it would be a prototype for a space colony. Eight adventurous people entered the two-year experiment on September 26, 1991. Biosphere 2 contained seven biomes: a tropical rainforest, savannah, marsh, 25-foot deep ocean, desert, an intensive agriculture area and a human habitat. I found the book *Biosphere 2: The Human Experiment* at a used bookstore. This book outlined the plans for each area. I was particularly interested in the intensive agriculture plans. When the crew sealed themselves in for the duration, they had some food that had been grown there in the planning stages and the crops were in full production. Their diet—which included meat, dairy, and eggs—had to be carefully planned. They had chosen a breed of chickens for eggs and meat. It was thought that these chickens would brood and raise their own chicks. The pygmy goats would be efficient eaters of the fodder crops they were growing to feed them. The small feral pigs would mostly get leftovers, starchy vegetables, and crop roughages. Worms and sorghum were planned to feed the chickens. They also had a fish-rice-azolla ecological system. Their crops were grown using Biointensive methods. This book was written at the beginning of the experiment. It all sounded great and I wondered what happened next.

I located a copy of *Eating In: From the Field to the Kitchen in Biosphere 2* written by Sally Silverstone, who was the manager of the food systems. From there I discovered that they were lucky to get any eggs at all from their chickens, let alone more chicks. They had limited food for the chickens, but production improved the second year when they fed them crop residues from the threshing that contained more seeds. The pigs became a problem and the last two became food for the table at the end of the first year. At least they contributed to the Thanksgiving and Christmas feasts. The fish were not as productive as expected and the bees failed. They were able to make a little goat cheese that was a welcome addition to meals. All in all, they didn't starve, but they all lost weight. Their staples were beans, rice, wheat, sorghum, sweet potatoes, white potatoes, and peanuts. Peanuts were an important source of fat. They also had bananas and papayas that they learned to use in creative ways. The vegetables and herbs from their gardens added diversity to their diet. Birthdays and other holidays were feast days. They would save limited supplies of special foods for these celebrations. Too often we get caught up in our everyday lives and don't take the time to declare "feast" days as we should. Can you put on a feast from only food you've grown? It is an interesting challenge.

Homegrown Fridays

For the past several years I've challenged myself to eat only what I've grown on the Fridays in Lent. It seemed like an appropriate time and having to do it every week for seven weeks kept me on my toes. I call my project Homegrown Fridays and have written about it in my blog at HomeplaceEarth.wordpress.com. I find myself labeling some things to save for the Homegrown Fridays if I have them in short supply. March is a lean time to be eating from the garden, so I'm calling on our staple crops for the meals. Carrots, beets, and greens will be brought in fresh from the garden. Actually, the roots will most likely be harvested in early March, before they send up a seed stalk and become woody. That last harvest can be put in the refrigerator. Stored crops and foods dried in the summer in the solar dryers fill our meals. The Irish potatoes are gone by then, but I could have sweet potatoes, garlic, winter squash, grains (mostly corn), cowpeas, and maybe onions. I do a limited amount of canning, since it takes more fossil fuel, lots of human energy, and heats up the house; but I do have green beans and tomato products that I've canned. Raisins made from my grapes, popcorn popped without oil in a pan, and mead made from our fruit and honey, add interest to the day. Tea is brewed from herbs grown here. In 2012 I added Thai Red Roselle (a type of hibiscus) for tea, then had to learn what part to harvest and dry (the red calyxes, not the flowers).

As interesting as it is to grow all your food, don't put too much pressure on yourself when you are starting out. You can learn a lot from putting together a meal, or a day's worth of meals, using only what you've grown, even if it is for only one day. Declare a feast day with your friends and have them contribute. You could start with a potluck with everyone bringing a dish that includes at least one thing they've grown. Make up your own guidelines. By doing things like that, your thinking will begin to change and your food plans will evolve. Be ready to be flexible with the plans you are making.

Oils and Sweeteners

I mentioned in Chapter 3 that you can make oil from peanuts, sunflowers, hazelnuts, and pumpkin seeds. Black oilseed varieties of sunflowers, such as Peredovik, are what you want. For pumpkin the oilseeds are "naked" varieties, the ones without the coarse white hull. Three varieties I have found that fit that description are Styrian Hulless, Lady Godiva, and Kakai. As you might imagine, their descriptions designate them as intended to be grown for their seeds, for eating or for oil, but not for their flesh. Maybe your chickens will like the flesh. The average US yield for pumpkin seeds is one pound per 100 ft².

I have American hazelnuts (filberts) growing on the north border of my garden. If you are serious about growing hazelnuts you would probably want to grow the European varieties. Although they are more susceptible to blight, the nuts are larger. Oregon State University has bred some that are resistant to blight. My yield from hazelnuts pressed for oil is about 3⅓ tablespoons per 1 cup nuts.

I have gotten 4 tablespoons of oil from 1 cup of peanuts. If you are used to a diet that includes using cooking oil liberally, thinking in these terms is quite a change. The oil produced in your garden becomes so much more valuable to your meals and to your spirit. A cup of peanuts weighs about 5.5 ounces. At 7.2 pounds per 100 ft² (US average) that would be 5.2 cups of oil per 100 ft² of garden space. If you used 1 tablespoon a day of peanut oil, you would use 22.8 cups of oil each year, or 1.4 gallons. At 5.2 cups of oil produced per 100 ft², you would need 439 ft² of garden space to grow peanuts for cooking oil. If your peanut yield was less than 7.2 pounds per 100 ft², you would need even more space.

It is nice to have a sweetener in your diet, and even nicer to grow your own. I mentioned maple syrup in the previous chapter. You wouldn't be growing maple trees in your garden, but you could have them, climate permitting, on your property. You would put one to three taps in a tree to yield 5 to 15 gallons of sap. It takes 10 gallons of sap to boil down to 1 quart (4 cups) of maple syrup.

Although I haven't been successful getting much syrup from the sorghum I've grown, I've seen it done. My efforts to squeeze the stalks were with a clothes wringer. The successful efforts I've seen are with sorghum presses powered by large animals or gas engines. Occasionally an old sorghum press like that comes up for sale at a farm auction. Bitterroot Tool and Machine, maker of the GrainMaker grain mill, is developing a homestead-sized press that should be ready for sale by the time this book is released.[1] According to the Master Charts, it is possible to get 1 gallon of sorghum syrup from a 100 ft² planting. You would need 10 gallons of sap for that. You would also need the right variety for your climate and the right tool to do the squeezing. What little syrup I did get was delicious. September is the time to make sorghum syrup around here and I find myself busy with other projects at that time of year. I'll stick to growing sorghum for grain. I like to have some around to grind and use in place of wheat when my gluten-sensitive friends visit.

Honey is my choice for a homegrown sweetener. When I started keeping bees in 2007 I could identify with the new gardeners who would show up in my classes at the community college. I had so much to learn and now it was me asking the beginner questions. If you keep bees, or are even thinking of keeping bees, join a bee club and plug into that support system. Bee clubs are a perfect example of the community we need to be part of.

If your bees survive the winter—they don't always—you could get up to 50 pounds of honey a year from one hive. The first year they are building up the colony, so don't expect any honey until the second year. Once you know what you're doing, you can use your own bees to increase the number of hives you have. If you are successful and have more honey than you can use, it can be one of those things you share through gift, trade, or sale. The bees will forage for several miles—yet another example where community is important. Your bees will be at your neighbors and their bees will be paying you a visit. You will be able to find lists of plants with nectar sources. Add as many of these plants as you can to your permaculture homestead. Eventually your perennials will need to be divided—a good opportunity to share with your neighbors. It will benefit you when your bees are out foraging. The busiest time for a beekeeper is spring and early summer.

Keeping Records

Pre-planning is great, but in order to know just how you are doing, you should keep records of your crops as you go along. Ecology Action has Data Report and Summary Yield worksheets that the GROW BIOINTENSIVE teachers use for reporting. These are included in their publication *Booklet #30: GROW BIOINTENSIVE*[SM] *Sustainable Mini-Farming Certification Program for Teachers and Soil Test Stations*[2] which is available as a free download at growbiointensive.org/publications _main.html. When I am in the garden I write notes on everything I do in 3" × 5" notepads, the kind with the spiral on top. Those notes are later transferred to Data Report sheets (I've modified the Ecology Action form to suit my needs). I use the information from the data sheets to fill out the Summary Yield form at the end of the year. Make sure to wear clothes with generous pockets when you are working in the garden. That way you have a place to hold a pen and notepad so those items are always with you.

For all of my garden plantings I have a record of what was planted where and when, and what amendments went into those beds. That is done right on my garden map—or rather maps. I record all the crops on one map, in the beds where they will be planted for the coming year. That is my "proposed" map. I have two more maps that are filled in as the season progresses. On my "actual" map, I fill in the crops in the beds as they are planted, along with the actual planting dates. Another map is my "amendments" record. Each time I add anything to the beds, such as compost, sulfur, or alfalfa meal, I record it there in the appropriate beds. I only maintain the data report record sheets, with the yields, for the crops that I'm studying for Ecology Action teacher certification and specific others I'm interested in. Keeping yield records takes time, no doubt about it. It might be that just knowing how much area you have planted will suffice for you, and if it is enough, too much, or not enough for your family for the year. Having a good garden map filled in with what you have planted in each bed, with the planting times, and maybe the end of harvest, is a lot of information. If you have a tight rotation, with the next crop planted as soon as the bed is free, the planting time of the next crop coincides with the end of harvest of the previous crop. With that information, you could look back in your small notepads, which are your chronological records, and find out more details, such as variety, seed source, etc. The most basic of records is your proposed garden map, as well as your actual and amendments garden maps. I consider those a must.

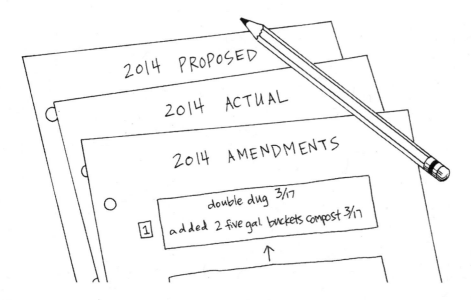

You could weigh or count each harvest, but that takes the fun out of just eating from the garden. Do that for those crops you are studying and enjoy the rest. It is helpful, however, if you know how much you have put up for the year. Keep a record of how many pints or quarts you may have canned of each thing or how many pounds of potatoes you put away. Making a note on your kitchen calendar might be enough to check back with later. What you have to put on your table will tell you if you've grown enough. If not, it is good to have something to look back on to help plan for next year.

5

Cover Crops and Compost—Planning for Sustainability

GROWING A SUSTAINABLE DIET also feeds the soil. Growing cover crops is one of the best things you can do for your soil. There is no separation between you and the earth that grows your food. Too many people feel distanced from where their food is grown, even though they aren't actually separated. Everything is connected and everything is important. From my study of nutrition and of the soil, I realized that the same thing is happening in the soil that is happening in our gut. Just as we have to make sure our bodies have the necessary nutrients and a healthy digestive system to assimilate those nutrients, we have to think of the soil in the same way. We need to eat food that is nutritious. It can't be that way unless the soil has a healthy digestion system and is full of nutrition. Healthy soil produces healthy plants, which feed healthy people, who populate healthy communities, which create a healthy world. Cover crops are food for the soil. They gather what they can from the sun, the rain, the air, and the soil they are grown in, turning it all into food that will feed back the soil with the plants as they decompose, both the above ground parts that you see and the extensive root systems that you don't see. It is the circle of life. One thing nourishes the next.

When I began gardening, the only information I had about cover crops was accompanied by information about the right time to turn them under with a tiller. Since I didn't have a tiller, I didn't consider cover crops, opting instead to cover my garden with leaves each winter. Using leaves is still a good idea, but not everyone has leaves available. As your garden gets bigger, as mine did, you need more and more leaves. Since my son owned a lawn service and one of his services was leaf removal, I had plenty of leaves available to me. There is a limit, however, to how many leaves one person can haul around on their garden, or make that how many leaves one person *wants* to haul around. The fourth year I was growing to sell to restaurants I bought a tiller and expanded my garden area. I only tilled once, and at the most twice, during the season. I maintained grass paths by mowing and treated each bed individually, not tilling the whole garden at one time. The beds in the market garden were 4' × 75'. I didn't use the tiller in the smaller gardens, which are the ones you see in my videos. Those were still covered with leaves for the winter. When I learned that I didn't need a tiller to manage cover crops I made the change to cover crops for all the gardens. Besides, I was beginning to worry that one day I might not have all those leaves available to me. What if Jarod decided to do something else and was not able to bring his leaves here? If my garden program depended on them, what would I do? I could find another source, but I wouldn't have the same confidence that they would bring me only leaves I could be sure weren't contaminated with anything harmful.

Beware of Bringing in Outside Inputs

About the time I was thinking about that, I began to hear about a new danger that had been ushered in with the new century. Through the twentieth century organic gardeners gathered carbon materials for mulch and compost from whatever sources they could find. If these materials weren't grown organically it was thought that the composting action would break things down and life would be good. In 2001 I learned about a new class of chemicals that were being used in the landscape industry and in agriculture that survived the composting process. These herbicides were used to kill broadleaf weeds in landscapes and to insure weed-free hay and grain in agriculture. If you used the resulting grass clippings, hay, straw, or even the compost made from the manure from animals that had eaten that hay or been bedded with that straw on your garden, your crops could suffer herbicide damage! Furthermore, the damage could persist for several years. I won't go into specific names of herbicides. As soon as one is banned another will be on the market. You could find more about this by searching "killer compost" on the internet. An early example of this problem that I read about was Penn State University using compost made with leaves from their own grounds. Even the leaves weren't safe anymore, unless you were sure where they'd been and what was used on the surrounding landscape.

Another consideration for not bringing mulch and compost materials from outside sources into your garden is what is happening to the soil where those materials were grown. Is that soil being depleted to feed your garden? Straw is part of the grain harvest. After the grain harvest, the straw could be returned to the earth. Otherwise, how is that soil being fed? How the soil is fed where your mulch and compost materials came from must be considered part of the ecological footprint of your garden, also.

Considering manure as a garden amendment, in a perfect world it would fertilize the land that grew the food for the animals that produced the manure. Ideally our manure would go back to the ground that feeds us. There are things to consider before you can do that wisely. If you are interested in learning more about recycling human waste, you'll want to take a look at *The Humanure Handbook* written by Joe Jenkins. Before the new chemicals were a problem, I used to occasionally bring in manure for fertilizer if it was convenient. One time I agreed to take the manure that had been building up at a horse owner's farm. About the time we had the last of it unloaded at my place, the horse owner mentioned she'd been spraying the manure piles with insecticide to keep the fly population down. That was the last time I hauled in any manure. If you are running chicken tractors (portable chicken pens) over your garden, their manure is fertilizer, but it is actually the grain you may have bought, that the chickens ate and pooped out, that must be considered when tallying up your sustainability.

You need to have your soil tested. Address any deficiencies with organic amendments as you establish your soil-building plan using cover crops. Whether you have clay or sandy soil, growing cover crops and using them in your garden will build organic matter right there.

Grow Your Own Compost

I was already studying GROW BIOINTENSIVE methods (which I'll refer to as Biointensive) by the time I learned of those new herbicides. Compost is a major tenet of this gardening method and it is made from crops grown in the garden just for that purpose. Compost crops are cover crops grown for compost making. The idea of growing enough cover crops to make all my compost was intriguing. With this method the cover crops are cut at maturity, or almost at maturity, and used in the compost piles. Too often gardeners don't realize the importance of using compost regularly in their gardens. It is not a product that is readily manufactured by those selling you things to start gardening. What you do hear about compost is making your own using your kitchen scraps. Plenty of companies will sell you compost bins to put them in. You would still need a source of carbon, such as leaves. The kitchen scraps are the nitrogen component. Unless you have a really big family and throw away a lot of food, you won't have enough kitchen scraps to make the compost you need. A compost with a ratio of 30/1 carbon to nitrogen is generally considered a desirable goal. To have a 30/1 compost, your pile would need to be made with an equal amount, by volume, of carbon and nitrogen materials. Biointensive compost has soil added, about 10 percent by volume of the built pile.

A recipe for making a 30/1 compost is: for every 2 five-gallon buckets of carbon material, you would need 2 five-gallon buckets of nitrogen (green) material, plus half of a five-gallon bucket of soil. (I take soil for this from a garden bed that is being prepared for the next crop.) Since I need to keep detailed records to maintain my certification as an intermediate-level Biointensive teacher, I record everything that goes in and out of the compost piles in my research areas. You can see that garden and watch me make compost that way in my DVD *Cover Crops and Compost Crops IN Your Garden*. You might have what I refer to as wild piles. Those are compost piles made as the summer progresses with anything and everything you have in the garden to throw in there. That's okay (I have some of those, too)—you pile it up and it eventually becomes compost. Once you begin to focus on what you are putting in there, and realize that you can grow things specifically for compost materials, your world begins to change in the garden.

If the ingredients in your compost piles are mostly leaves and other carbon materials, the pile will have a higher carbon-to-nitrogen ratio and take longer to break down. That's not a bad thing, you just have to be aware of it. If it has more green nitrogenous material, it will get hot fast and might smell. Did you ever leave some grass clippings in your wheelbarrow, forgetting to put them somewhere as mulch or add them to your compost? A pile of damp grass clippings, left for a few days, can stink worse than any manure. A compost pile built with a higher concentration of green material is a hot compost. It breaks down faster, but you don't have the diversity of microbes that you do with cooler, slower compost. The compost I'm talking about making is the cool, slow kind. My compost piles are made on a garden bed and are part of my rotation. The piles I make spring through summer only get turned in the fall as I move them to the next bed. Those piles will have become finished compost by the following summer when I add it to the beds as needed. The pile I make in the fall is made on the new compost bed and fed out to the garden the following fall. That pile is never turned. I try to make sure my compost piles have enough moisture; otherwise I leave them alone over the course of the year. Sometimes I put a post at each corner of the pile to keep its shape as I build it, and sometimes not. When the compost is removed from the garden bed in the fall, that bed is planted to rye. In the spring, the rye is cut down at pollen shed (I'll explain that soon) and left as mulch. As it is growing, the rye soaks up any nutrients leached into the soil from the compost over the year and gives them back as it decomposes under the corn that is planted into that mulch. My best corn crop is in that bed following the compost.

Sixty Percent Compost Crops

To grow enough compost materials yourself, you would need to have those crops growing in at least sixty percent of your garden over the course of the year. Of that sixty percent area, two thirds would be in carbon crops and one third grown for their nitrogen contribution. Too often, gardens are productive for only part of the year and left to whatever Mother Nature wants to put there for the rest. If Mother Nature's choice is not yours, you will have to expend energy getting it out. Once you incorporate cover/compost crops into your garden rotation, and understand the process, everything will just be a matter of planting and harvesting. Your harvests of green biomass and brown stalks and straw are food for the compost pile. Properly managed, you don't need to till before planting again. Keep in mind I said *once you understand the process.*

Some of these crops will already be in your food plan. The stalks of corn, sunflowers, and Jerusalem artichokes, and the straw from rye, wheat, and other grains, are your carbon sources. Using a machete, the stalks can be cut and chopped into lengths suitable for composting. I use a Japanese-style sickle to cut the small grains. You'll find photos of these tools in the color section of this book. You need to keep the sickle sharp for the job and remember to wear gloves. Accidents can happen in an instant. Just ask me how I know.

The nitrogen contributing crops are legumes grown not for food but to provide biomass for the compost. Some of these crops; such as hairy vetch, Austrian winter peas, and crimson clover; are planted in the fall and grown through the winter. Many gardeners might already be familiar with them. Fava beans are winter hardy in some areas. Summer crops would be cowpeas and buckwheat. Most of my cowpea plantings are grown out for dried seed for us to eat, but if I needed biomass for the compost from cowpeas I would grow it just for compost and cut it when it is flowering. That's when it has reached its most mass. After that the plant energy goes into making seed. Just as with the small grains, I use the sickle to cut the legumes. I'm working in four-foot wide beds. If I was cutting a larger area I would use a scythe.

Legumes can be planted with the small grains (wheat and rye). "The mixture may produce more biomass, form more soil organic matter and suppress weeds more effectively than either component alone."[1] The more diverse habitat of the combination will also attract a more diverse community of beneficial insects and soil microorganisms. You only need a small amount of legume seed added to the grain when planting, or else the legume will overwhelm the grain. It is not so much of a concern if you've planted a legume with rye and you are cutting it at pollen shed for mulch. However, if you are growing the grain out to save the seeds too much legume will pull down the grain plants, diminishing your grain harvest. Also, by that time, the legume would have set seed itself. The legumes I use for this combination are either Austrian winter peas or hairy vetch. If I'm growing wheat and rye to seed, I pull out the legume plants when they are flowering. Austrian winter peas are easier to do that with than hairy vetch. As a result, I add only Austrian winter peas to the grain beds destined for a seed harvest. Hairy vetch is a bit more wild and vining, tangling itself with the grain plants, which is okay if it is cut and left as a mulch when it is flowering.

Red clover (different from crimson clover) and alfalfa are legumes that will be in your garden for more than one year. Red clover can be planted in early spring one year and grow through the next year before another crop replaces it. I get one cutting the first year and two the second year. Alfalfa could stay in even longer than that. Red clover is shown in a rotation on the Garden of Ideas Map (Figure 8.4). These crops take a little more skill to work into your garden rotation than the other legumes I've mentioned, but they will reward you with more nutrient-packed material.

Buckwheat is not a legume, but it is a good filler in the summer, flowering in 30 days. It crowds out weeds, scavenges phosphorus in the soil, and the bees and other beneficial insects love it. However, it doesn't offer much biomass. Keep some handy for when you find there is a space open for a few weeks between crops, but don't depend on it for much compost material. To find which cover crops grow best in your area inquire at your state's Cooperative Extension Service. It used to be that we had to stop by the local office to pick up their publications, but now everything is online. One of the best books to have as a cover crop reference is the third edition of *Managing Cover Crops Profitably*.

To help in determining if I have at least 60 percent of my garden in soil building compost crops, I use a worksheet (Figure 5.1) to determine Bed Crop Months (BCM). Each garden bed has 12 BCM for the year. If you were to have a compost crop in one bed for 60 percent of the year it would be in there for 7.2 months (or 7.2 BCM), leaving 4.8 BCM for other crops. In practice some beds may have compost crops in for a whole year, such as a bed with winter rye followed by corn followed by winter peas. Some may not have any at all if they are filled with summer vegetables and garlic or greens for the winter. Everything gets rotated, so you don't have to think of the individual beds themselves having the 60 percent. Your target will be at least 60 percent of the BCM of the whole garden for the year. Assuming your beds are the same size, multiply the number of beds you have by 12 to find the total BCM. If your garden has ten beds, you will have 120 BCM. Sixty percent of that is seventy-two.

When I use the worksheet I have my "proposed" garden map as a guide and, beginning with one bed, list every crop that is in there for the year. Continue for each bed. In the appropriate column (compost crops in 60%, other crops in 40%) multiply the number of beds times the months that crop is in the bed. Most often the number of beds is one, but sometimes it is a partial bed, so that number is a decimal. I might have corn and sweet potatoes planted in the same bed. The corn would be 0.5 beds in the 60% column and the sweet potatoes would be 0.5 beds in the 40% column.

I go by the calendar year. A compost crop that would have been planted in the fall and harvested in mid-June would count as 5.5 months. Those months from planting through December 31 would show up on the previous year's worksheet that goes from January through December. In Chapter 8 you will find examples of garden maps with all the crops filled in. It might be that you haven't determined all your crops yet and are using this worksheet just to see how things look so far.

Date: Garden BCM:

BED CROP MONTHS

Figure 5.1.
60/40 Bed
Crop Months

Bed #	Crop	60% Crops BCM # of beds X months in bed=BCM	40% Crops BCM # of beds X months in bed=BCM
		Total BCM	
Total BCM/Garden BCM = % of Total			

Homeplace Earth
Education and Design for a Sustainable World
www.HomeplaceEarth.com

Download this worksheet at http://tinyurl.com/mf4a33r

When you have all your crops listed and in the appropriate columns, and you have determined the BCM, total the columns. Divide each column total by the number of BCM for your garden and you will have the percentage of compost crops and of your other crops. If you are not at your target, don't despair. You know what you have to work on. If this is new to you, it will take some time to make the transition. Eventually you can fine-tune the compost crops so two thirds of the 60 percent BCM are carbon crops and one third is legumes.

Using the term Bed Crop Months works if all your garden beds are the same size. If they aren't, you might prefer to do your calculations using the amount of square feet per crop. Multiply the planted area of your garden (in square feet) by 12 (months). We'll call that number **A** (for area). On the worksheet, instead of the number of beds that is planted to a crop, you would put the number of square feet that is planted for that crop. At the end, divide the total square feet for each column by **A** to find the percentage.

Cutting Rye at Pollen Shed

I've found that cereal rye, sometimes known as winter rye, gives me more straw than wheat does to use as a carbon material in the compost pile. Whatever you see on top of the ground, rest assured that there is at least that much organic matter below ground that will decompose and feed back the soil. Just cut the top off and leave the roots there. There are some places in my garden plan that cutting the rye at pollen shed is the best choice for me. Pollen shed is the point at which the plant is flowering. You will find a photo of rye shedding pollen in the color section. Once it has reached that point it is most likely that if you cut it near the ground it won't grow back. You can cut it and just let it lie on the bed, providing mulch to the next crop. Since the rye was actively growing up to that point, it will take a couple weeks for the roots to loosen their grip, so wait until then to transplant into that bed. At this point the soil is suitable for transplants but not seeds. In my area in Zone 7 the rye is shedding pollen about May. Waiting two weeks, the next crop would be transplanted into that bed around May. If you wait too long after pollen shed to cut the rye, viable seeds will have formed, something you don't want in your mulch. Unseasonably warm winter temperatures, such as we had in 2012, will alter the timing.

Harvesting the Grain

If the goal is to harvest mature grain and straw from wheat and rye, it would be cut when the grain is ready—about the middle of June at my place. Wheat is ready here a week before the rye. I cut these grain crops close to the ground with the sickle and gather some together into bundles. Using a fairly green straw from what I cut, I wrap the straw around each bundle and tie it. If that doesn't work for you, you could use baling twine for tying the bundles. I make a shock out of about five bundles, standing them up and leaning them against each other. The shocks stay in the garden for about a week until the grain is threshed. Nightly dew and an occasional rain won't hurt the grain. You could store the bundles in a barn until threshing if you can protect the grain from mice.

I have two methods of threshing. One requires an old bed sheet, a piece of plywood at least 2' × 6', something to lean the plywood against (I use the picnic table), and a plastic baseball bat. Lay the sheet on the ground to catch the grain. Put one end of the plywood on the sheet and lean the other end against the picnic table or whatever you are using to support it. Working with one bundle at a time, hold the wheat or rye by the straw on the plywood, grain heads pointing down, and hit the grain heads with the bat. The grain will separate and roll right down the plywood to the sheet. There will be bits of straw, chaff, and grain in the sheet. I put it all through a piece of half-inch hardware cloth to take out any straw. The next step is to winnow the grain.

Traditionally, the grain and chaff was tossed in the air. The wind would blow the chaff away and the grain would fall into a clean container. Since I can't count on windy days when I thresh, I pour the grain and chaff from one bucket to the next in front of a fan to remove the chaff. Wanting to have clean grain to put away in my pantry, I wash it next. To do that I put small batches at a time in a very large bowl filled with water. The chaff, any other dirt, bird droppings, and small insects float to the top and can be poured off. I'll fill the bowl with water again until there is nothing to pour off but clear water. The grain is then drained and spread out on towels in the house until dry. This method of cleaning the grain might not be suitable if I had bushels and bushels, but for a household amount that comes from the garden, it works for us.

Rather than threshing with a bat, an even easier method is threshing with your feet. The sheet is spread on the ground as before, and this time I lean a frame with half-inch hardware cloth attached to it against the picnic table bench. Sitting in a chair in front of that frame and wearing a clean pair of sandals that I keep for that purpose, I hold the straw with the grain heads against the hardware cloth and proceed to shuffle my feet back and forth. The grain falls through the half-inch wire onto the sheet. Winnowing is the same as before.

Even if you aren't growing grain to eat, you can grow it for the straw for compost material and you will have seed to save for your next cover crop at the same time. Since the grain crop has finished its growth cycle, the bed is easily prepared for the next crop, leaving the roots in place. Without removing the stubble or doing any tilling, transplants can go in or rows can be easily hoed in the stubble for planting seeds. Although I usually broadcast the grain in the fall to plant it, in the area where I'll be planting carrots and beets the next summer, I seed the rye in rows. When the grain is harvested in June, there will be rows of stubble in the bed. I lightly hoe between the stubble rows and sow the seeds for the root crops where I've hoed. The stubble protects the new roots as they begin to grow and feeds them as it slowly decomposes. I pay particular attention to watering that bed after planting to get it off to a good start and reseed if there are spots with poor germination. After things are up and growing little extra attention is needed until I harvest carrots and beets for fall and winter food. In the spring that bed would be empty and ready to plant to a new crop.

If you have children or grandchildren around for the harvest, growing your own grain is fun to do and what a learning experience they will have. Another fun thing to do around grain harvest time is to go to the library and find as many different versions of *The Little Red Hen* as you can. I did that one year with my grandson. We were surprised how many different stories on the same theme there were—including one set in the city with the wheat being made into pizza!

I sold my tiller years ago and am happy it no longer takes up space in the barn. Some of you, no doubt, will be using mechanized equipment on your land. Pam Dawling, garden manager at Twin Oaks Community in Louisa, Virginia, uses cover crops, often managing them using tractors and tillers. Whether you are using tractors, tillers or hand tools, you will find much helpful information about cover crops and growing your food in general, in her book *Sustainable Market Farming*.

6

Companion Planting

OVER THE YEARS I've heard many people say that if they can't eat it, they're not going to grow it. To be honest, in my early years of gardening I was that way also. However, we would do better to take a broader perspective. Our food crops are part of a much bigger ecosystem that all works together and we have to consider the whole of it. That's where companion planting comes in. It can mean planting two or more crops next to each other or planting them nearby, such as in a border. When crops are mixed in the bed it is called interplanting. When plant-eating insects are scouting out their targets they are attracted to monocrop systems where there are whole fields of one thing—all their favorite food in one place, how nice for them! A sustainable garden will be developed with diverse plantings that attract beneficial insects, that will then feed on the not-so-beneficial ones, all without chemicals. These plantings will enhance the beauty of your garden and increase the flavor of your vegetables.

I'm not going to give you lists of what grows well together and what doesn't, although I'll be giving you some examples. There are plenty of resources where you can find that information, including *How to Grow More Vegetables* and *Carrots Love Tomatoes*. What I am going to tell you about is my experiences of seeing what happens when an ecosystem is working. You can worry yourself silly trying to get the "right" combination of crops together and avoiding the "wrong" ones. It will be much easier if you understand the habits and life cycles of both the insects and the plants.

Some plants, particularly those with small flowers, are good at attracting beneficial insects. The beneficial ones are those attracted to the nectar and pollen of the flowers, but that also feed on the larvae of the insects you want to have less of in your garden. So, leaving some things to flower and go to seed in your garden is what you should strive for. When I first grew basil I had read in the herb book that you need to harvest it before it flowers for the best culinary use. I thought I was not being a good and attentive gardener if it went to seed. Well, if you've ever grown basil, you know that it is hard to keep trimmed all the time and eventually, some of it will flower. One year, the basil had gotten ahead of me and I went to the garden with my clippers to get it back under control. What I found was that my basil was teaming with life. There were so many insects, including tiny wasps, buzzing around those plants it was amazing! They were after the flowers. The insects being attracted were beneficial ones. I eventually trimmed the basil, but not that day. The best time of day to witness something like that in your garden is between 10am and 2pm. Now I make sure to leave some of the basil flower every year. Often the best crops to grow together are the ones you eat together and so, for example, I plant basil with my tomatoes.

Dill goes well with squash and cucumbers. All the umbelliferous plants—dill, carrots, parsley, and celery—are good plants to have. Celery, parsley, and carrots are biennials, flowering the second year. If you leave them to overwinter in your garden, they will be up early in the spring and flowering without you even needing to think about it. If you like to use celery seed in your cooking, just watch for the seed to mature and gather it. You will have enhanced the beauty of your garden in early spring, attracted beneficial insects, and produced food for your kitchen—all while just letting nature take its course.

Not enough gardeners and farmers save seeds. The value of doing so is that the process of saving seeds sets the stage to provide food and habitat for the "good" bugs. I'll be talking more about seeds in Chapter 9.

I've heard of people buying ladybugs and other insects to put in their garden. You could do that, but if there is not enough food and habitat for them they won't stick around. It is sort of a build-it-and-they-will-come deal. If you provide the right habitat, they will show up on their own. I began growing cowpeas because they were drought tolerant and we were in the midst of some drought years at the time. Cowpeas have become our staple crop for dry beans. I could usually find ladybugs on the cowpea flowers and one day when I was in the garden with the camera I decided to photograph one. The ladybug I had in mind was scurrying up and down the plant stems and I had to wait until she stopped. When she did, I snapped the picture. Upon a closer look, she had stopped to munch on an aphid. I didn't even know about that aphid and it was being taken care of. You can find that photo in the color section of this book. In a sustainable garden the goal is to have a balance, not to eradicate any one thing.

Many years ago our son, Luke, was helping me in the garden and found a really strange looking bug. At the age of twelve he knew that you don't just go around killing bugs—at least not until you've identified whether they are harming your crops or not. A university professor had been coming out each week to do some studies on our farm, so Luke saved it to show her. She identified it as a wheel bug, also known as an assassin bug, and explained that it is a beneficial insect that will suck the guts out of caterpillars. Luke set up a terrarium and kept it to study for about two weeks before he let it go back to the garden. Meanwhile, when we were working in the garden or washing lettuce for the market and came across a little caterpillar, he would feed it to his new pet. Sure enough, it sucked the guts out as he watched! Nature in action. That wheel bug wouldn't have been there if we hadn't provided an attractive home—a welcoming habitat. The professor coming to our farm thought it was wonderful to see so many insects here. I took it for granted and suggested that all farms had insects like that. She assured me that conventional farms did not.

A guide I've used often to find out which insects are good to counteract the undesirable insects, and what to plant to provide the necessary habitat, is *Farmscaping to Enhance Biological Control*. It is a publication from ATTRA, which is a project of the National Center for Appropriate Technology. Much helpful information is available on the ATTRA website, including many short videos and longer webinars about sustainable agriculture. Although the ATTRA information used to be free, there may now be a charge for some of the publications due to federal budget cuts.

Farmscaping is a word coined by entomologist Dr. Robert Bugg to describe a holistic approach to insect management. The field days I attended on no-till farming methods were also about farmscaping. The publication that resulted from that research at Virginia Tech is *Farmscaping Techniques for Managing Insect Pests.* One of the authors, Dr. Richard McDonald, is the best presenter I've experienced on the subject of farmscaping. Both in the field and in the conference hall you can sense his excitement for insects. You will find lots of information and photos on his website at drmcbug.com.

Potatoes

When we first moved to our five acres in the spring of 1984 I planted a garden and put in potatoes. I'd had a small garden where we lived before and had grown potatoes with no problems. We got busy fixing up the house and the next thing I knew the potatoes had grown up and were covered with Colorado potato beetle larvae. There were few gardens where we lived before and I was probably the only one with potatoes there. Here, plenty of people grew potatoes and much bigger patches than mine. I asked around and the acceptable approach, at least for my neighbors, was to spray the insecticide Sevin to combat the potato beetles. I even asked a friend who I thought adhered to organic practices and she also suggested Sevin. That was not an acceptable approach for me. If I wanted potatoes treated with insecticides I might as well just buy the ones in the store. I taught our children, who were turning 11, 7, and 3 (all have summer birthdays), to identify and pick off the larvae and bugs. We did that for the next few years as part of our garden chores. My problem with potato beetles lessened and then just disappeared. I had developed the ecosystem so that things were in balance. I can't say exactly what I did that ended my problem with the potato beetles. The tansy I got started would have helped, but that was really only part of the whole.

In 2002 I was able to plant a garden at the community college where I was teaching sustainable agriculture classes. I put in potatoes and was disappointed that they were plagued by Colorado potato beetles. I had not had to think of those insects in years. I checked the ATTRA farm-scaping publication and saw that the spined soldier bug targets potato beetles and the way to attract them is with the sunflower family (golden-rod and yarrow), bishop's weed, and maintaining permanent plantings. I did not check into bishop's weed at the time, but a quick internet search now indicates that although it is good for filling in shady areas, it can become pretty invasive. I sometimes had sunflowers in that garden, but they would have been blooming too late for the potatoes. The rest of the college garden contained annual plants, which were not much help in providing the perennial plantings that the publication indicated would be useful. I did put in some tansy along the fence surrounding the garden, but it wasn't enough compared to what was going on all around. Herbicides were also in use. How the area surrounding the garden was managed had an effect on the garden itself. It is in building the ecosystems that we regain balance, not spraying chemicals.

Achieving Balance

Anytime you change something it upsets whatever balance was already established. Even if you are doing all the right organic things, when you put in a garden where there wasn't one, the balance that was there needs to adjust to the new plants and system. A farmer needs to have not used chemicals and to have done soil building practices for at least three years before organic certification is awarded. It takes three years for that new balance to be reached and for things to be working as they should in the system. During the years I was selling produce I had two friends who had started farms tell me that they were doing all the good things they should be doing and they were not yet seeing results. They didn't quite know what it was that they should be seeing, but they knew it wasn't there. They were each probably beginning their third year of growing at the time. I told them that they hadn't been doing it for the full three years yet, and if they kept at it they would see the change. I am fortunate that they both came back to me later and said that it had happened on their farms, just as I told them it would. They could see and feel things coming into balance. If you are used to taking the fast and easy approach, it can be hard to wait. If you have a partner who prefers reaching for the chemicals over watching what will happen, it is even harder. Even using a little herbicide or insecticide sets things back. You can't have it both ways. Look for organic controls that you can use while your system is developing.

The first year I became a market gardener my main crop was lettuce. I had grown lettuce for my family before, but that year it was in a third of my garden. Talk about upsetting the balance! Things were going along fine until a week before my first harvest when the slugs moved in. I discovered that the slugs came out on the leaves just before dusk and were still there in the early morning while the dew was on the plants. I would go out with a spoon and cup to scrape them off. It was only a problem until the hot days of summer set in. I had to be diligent the first year. The second year I was watchful and kept after the slugs for the few weeks they were a problem, but they never were as bad as the first year. After that I don't remember ever having a problem. We would see slugs when we washed the lettuce for market, but they didn't do enough damage to think about. Since we had a healthy ecosystem in the garden, I imagine that the birds and toads worked slug harvest into their routine and took care of them for us.

I mentioned that when you mix different crops in the same bed it is called interplanting. In permaculture, planting certain things together like that is referred to as a guild. Sometimes I have interplanted things for convenience, such as lettuce and cabbage. I set the cabbage out at their regular spacing—about 15"—and the lettuce transplants go in between the cabbage plants. They all grow together in the bed until the leaves begin to touch. That's when I harvest the lettuce. With lettuce in the bed there is no space for weeds. If you find things that work well together, think of them as a guild that is always planted together. The Three

Sisters—corn, beans, and squash—is an example of interplanting. Sally Jean Cunningham wrote *Great Garden Companions*, a good resource for information on companion planting. She mixes herbs, flowers, and vegetables in her beds. That book has been an inspiration to Brent, one of my friends you meet in my garden planning DVD. He has moved since we filmed for the video, but only to another part of Richmond, Virginia. His front and back yards are filled with gardens and he even has chickens now. Some of his beds look like they could come right out of Cunningham's book. Once when I was there I saw white cabbage butterflies flitting about his yard, but no holes in the leaves of his brassica plants. I also saw wasps flying along those greens looking under the leaves for a snack. I'm sure they were looking for larvae or eggs. Everything was as it should be. Although I don't remember all that he had planted with those cabbage family plants, I do remember that hairy vetch was blooming in there.

Borders

Developing borders is a great opportunity to put plants together in the garden, while establishing permanent plantings and habitat. Even annuals will do to start. Here's where you can put the herbs and flowers that add so much to life. You can plant bulbs in your borders as well. This is all part of establishing the ecosystem. Think of it as a mini-hedge. In fact, it is good to learn about hedges. They are useful wherever you have room for them. One side of my large garden has always had a fence, so the honeysuckle, day lilies, and other things that just grow there are permanent plantings. It's sort of wild. The other three sides are planted on the inside of the fence, but most years grass grows up to the fence on the outside. Big on my to-do list is to establish a perennial border on the outside of the fence on those three sides. On the inside, I have established hazelnuts (filberts) on the north side of the garden and blueberries on the west side. Annual vegetables are on the south side.

One summer I had some great looking marigolds growing on the outside of the garden fence. Marigolds are always good to have in and around your garden. Our son Luke had a team of calves that he was training as oxen and they had reached the limits of our pasture. For some extra grazing, he put up electric fence to use the grassy area surrounding the garden. We assumed the 4' garden fence would suffice to keep them from the garden. I thought I would have to sacrifice the marigolds, but the steers didn't even nibble on them. Since they stayed away from the marigolds, they stayed away from the fence. The following spring, they were let into that area again, but there were no marigolds along the fence. The steers came up to the fence and dropped their heads over, leaning on the fence and eating the collards that I had going to seed. Granted, the animals were bigger now, but that made me think that if I had highly scented perennials there, maybe those plants would keep back the steers and possibly other animals. The marigolds were great for that, but they are seasonal. There are books about plants that wildlife don't like. Maybe a combination of those plants, along with a fence, would mean the fence doesn't have to be so high. Luke's oxen have moved on to other pastures, but now that we've considered that area outside the garden as potential extra grazing, I want to establish a border that won't get eaten. It will be an interesting challenge.

Comfrey is good to have in your garden. A border of comfrey planted eighteen inches apart has kept wire grass from encroaching in one of my gardens. Comfrey dies back in the winter, but pops up in early spring to form a nice green fence. You could harvest comfrey for compost material or to feed small livestock. In that case, their manure goes to the compost. Comfrey won't do, however, in the border that I don't want grazed by the steers. They ate it down so thoroughly along one fence that it never came back.

I discovered tansy in my early years here at the farm when I was searching for a solution to the beetle problem with my potatoes. I had read that tansy was a good companion plant for potatoes, so I planted tansy seeds in my potato bed. The tansy grew well, but of course the new plants didn't do much for the potatoes that year. Since tansy is a perennial, it wouldn't be good to leave it in the garden bed, so I transplanted it along our porch. It has done great there. It is still close enough to the garden so the beneficial insects it attracts are nearby and, since it tends to spread out, I can dig some at any time to plant another place or give to friends, redistributing the surplus. Cunningham refers to tansy as "probably the single best attractor for beneficial insects." The list of insects she's found on her tansy includes the spined soldier bug, the same one the ATTRA Farmscaping publication said would go after the Colorado potato beetle.

How the paths are handled is a consideration as part of your companion planting/farmscaping plan. Tilling destroys any habitat that might be there for the beneficial insects. In Chapter 2 I mentioned that I have white clover or leaf mulch in my narrow paths and grass in my wide paths. This provides a nice habitat for beneficials. According to the ATTRA Farmscaping publication, ground beetles like to live in permanent plantings, white clover, and mulch. They may help with the Colorado potato beetle problem, in addition to problems with slugs, snails, cutworms, and cabbage-root maggots. Just imagine if chemicals are sprayed in the paths. Besides the vegetation, all the beneficial life in the paths would perish.

Don't be too quick to clean up the edges of your property. Eradicating every weed and trimming every blade of grass is considered desirable to some, but at what cost? Weedy fencerows, as well as other shady, moist spots can be homes to spiders. Spiders eat only insects, and live ones at that. All those insects get thirsty, so make sure to provide some water for them. It can be in shallow dishes you set out, a pond in your garden, or even a small wetland. I have only touched the surface of all you could do with companion planting, but I hope I've given you an understanding of what is necessary to allow all the benefits of an ecosystem to develop. Diversity is what we are after in a sustainable garden. The more different things we have, the better.

7

Plan for Food When You Want It

Ya Gotta Have a Plan! That was the working title of our garden plan DVD. The final title is *Develop a Sustainable Vegetable Garden Plan*, which is more descriptive, but I like the working title better. "Ya gotta have a plan" was what I found myself saying to people who would come to me with questions. It would be nice if all you had to do was walk out to your garden and spirit would guide you on what and when to plant, and when and how long your harvest would be—and that might very well work for you. I partner with spirit a lot and know that spirit can't do it alone. Being in a partnership means we have to do our part, which means do our homework by learning all we can about what we want to eat and what we will be planting. For those of you who discount spirit in your lives, all the more reason to do your homework, since it's all on you. Have you ever heard that you need to develop clear intentions of what you want to accomplish? A plan is where you lay out your intentions for the garden for the year. Things don't always go as planned, but that is part of the experience. Everything becomes a learning experience, whether it worked out as you expected or not. Stay flexible and don't ever stop learning.

There are a lot of steps to planning and growing a sustainable diet. Sometimes you'll be working on many steps at the same time. No doubt, by this time you have a list of things you want to grow and are wondering how you will fit it all in your garden. If you've made your garden map, you know how much space you have to work with. Fitting everything in has a lot to do with timing.

Unless you've been growing for a while, you might not know how long it takes from seed to transplanting in the garden, how long that crop will be in the garden until the beginning of harvest, and how long the harvest might be. Even if you are an experienced grower, it is good to think it out anew once in a while. Having that information in one place is really helpful, which is why I've come up with worksheets. In this chapter you will find the Plant/Harvest Times (Figure 7.1) and the Plant/Harvest Schedule (Figure 7.2) worksheets. One of the great things about teaching all those years at the community college is that every year I went over all the material again, refining it each time. When I grew for my family, and then for the markets, I worked with this crop information without the benefit of worksheets. Having it all in one place facilitates planning and is helpful mid-season when times are busy. You don't have to stop and think just which catalog you found that seed in to check the days to maturity, or whatever it is you have a question about.

Frost Dates

Taking a look at the Plant/Harvest Times worksheet, you will see a place for the dates of the last spring frost and the first fall frost. Of course, you don't actually know when those dates are until they pass, but you can put the expected dates for your area there. If you don't know when that is, you could call your county Cooperative Extension Service, ask your gardening neighbors, or find the dates online at plantmaps.com. For starters, put the expected dates there, but later on you could go back and add the actual dates. You will have a record specific to your garden. A word of caution—just when you think you know what to expect, things change. (I know I already gave you an example of that in Chapter 3 with the rainy October, but here is another one.) Our usual last spring frost date is about April 25 here. We live out in the country.

In the city of Richmond, Virginia, just fifteen miles away, it might be a little earlier because of all the concrete soaking up the sun. I would put out my tomato transplants on April 25 and mulch them with leaves so that I wouldn't have to weed at all. One year the weather was warming up nicely and I put out the tomatoes, as usual. I got busy with family and life in general over the next week and obviously was not paying attention to the weather forecast because we had a killing frost on May 1 and I lost many tomato plants. I realized later that they might have survived if I had not mulched them. The soil was warmer than the air and if the mulch wasn't there, the soil would have probably warmed those

Figure 7.1.
Plant/Harvest Times

Bed #	Crop, Variety	Days to Maturity		B	C	D	E	F	G	H	I
		A1	A2	date to plant in flat or cold frame	weeks in flat or cold frame	date to plant in garden	weeks in ground until harvest	begin harvest date	weeks of harvest	end harvest date	Notes
		from seed	from transplant								

Last Spring Frost _____

First Fall Frost _____

_____
(year)

Homeplace Earth
Education and Design for a Sustainable World
www.HomeplaceEarth.com

Download this worksheet at http://tinyurl.com/mf4a33r

little transplants enough to keep them alive. With the mulch down, the transplants were insulated from the heat of the soil and left to the mercy of whatever temperature the air was.

My neighbor came to my rescue with tomato plants. She called me to say that she had planted out all she needed and still had plants if I wanted any, without knowing what had happened to mine. She had planted her tomato seeds in a coldframe on April 1, covered by a sheet of plastic—nothing elaborate—and they were the greenest, stockiest tomato transplants I had ever seen. I had planted mine inside under grow lights on March 1. I learned four lessons from that experience: (1) watch the weather, (2) don't be so quick with the mulch, (3) you don't need grow lights to have great tomato seedlings, and (4) just because someone might use Sevin on their potatoes doesn't mean they don't have good gardening advice (and plants) to offer. I haven't used grow lights for years now and start everything in coldframes. My tomato seeds go into their coldframe about the last week in March.

It is not only the date on the calendar but also the temperature of the soil that you should be watching. If you have mulched your garden beds over the winter, pull the mulch off a couple weeks before planting your spring crops to let the soil warm up. You can tell a lot by just putting your hand in the ground. A bed with a crop that has winterkilled, leaving the soil ready to plant, will be warmer than one that has had mulch on it all winter.

Usually our first frost in the fall is in late October and sometimes not until early November. When growing crops for fall and winter harvest, having a date to use for planning is important. With carrots and beets, I want to have them mature by then. After that, they can stay in the ground for harvest right through winter. Once the nights begin to cool and the days are shorter, plants grow a little slower. You may need to add as much as two weeks to the days to maturity time if you are planning crops that will mature in the fall. Although it is usually later, one year we had a killing frost on October 9. I use October 15 as my first frost planning date. I harvest sweet potatoes and peanuts before then so their foliage is not damaged by frost, all the better for compost material.

Top:
Garden in early
March.

Middle:
Same garden in June.

Bottom Left:
Rye shedding pollen
in early May.

Bottom Right:
Rye cut for mulch-in-
place at pollen shed.

Top:
Garden map and Plant /
Harvest Schedules.

Below:
Ladybug eating an aphid
on a cowpea plant.

Below Right:
Sunfield Farm
Permaculture Plan.

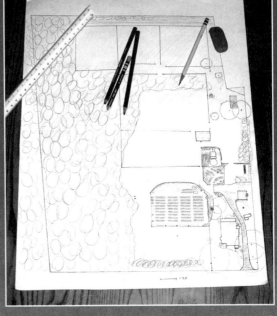

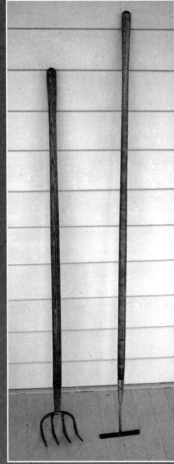

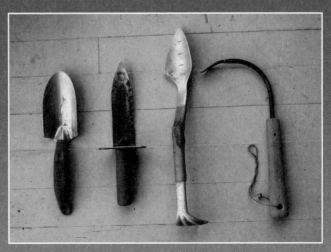

Top Left:
Spade, garden fork, mattock.

Top Right:
Cultivator and collinear hoe.

Bottom Left:
Sickle and machete.

Bottom Right:
Left to right: Trowel, Lesche soil knife, Trake, Cobrahead.

Top:
Ginseng sweet potato slips grown in a coldframe, ready to cut off for transplanting.

Middle:
Purple sweet potatoes with slips grown in jars of water.

Bottom:
Sweet potatoes freshly dug with a garden fork.

Top Left:
Grainmills left to right:
Grainmaker, Country
Living, Corona.

Top Right:
Threshing with the
plastic bat method.

Middle:
Threshing set-up for
the foot method.

Bottom:
Wheat (right) is ready
to harvest. Rye (left) will
be ready a week later.

Top:
Tomato juice (left) and tomato soup (right).

Middle:
Pressure canner (left) and water bath canner (right) with canning jars and two-piece lids.

Botttom Left:
Dill (sour) pickles fermenting in a gallon jar.

Bottom Right:
Traditional straight sided crock, hand carved stomper, Harsch crock.

Top Left:
Corn and bags of cowpeas hanging in the barn waiting for shelling .

Top Right:
Corn and cowpeas in the pantry.

Middle:
Mississippi Silver cowpeas and Bloody Butcher corn in the garden.

Botttom Left:
Corn sheller in action.

Top: Solar food dryers in the garden.

Lower Left: Inside of solar dryer based on SunWorks design.

Lower Right: Inside of solar dryer based on Appalachian State University design.

Plant/Harvest Times

Seed catalogs are a wealth of information. At the beginning of each crop listing you will find a box with information specific to that crop, such as when to plant, diseases, how many seeds per packet or ounce, and more. If you are ordering online, check the website for cultural notes. The specific varieties of the crops you choose would determine things such as days to maturity and length of harvest. The descriptions in the catalogs can be quite compelling. Unfortunately, we quickly forget just what was so special about the varieties we chose. If you are not opposed to cutting up your seed catalogs you could cut out the pictures and descriptions and put them in your garden notebook for a ready reference. Better yet, cut up last year's seed catalogs. If you've ordered online, you can print off the descriptions for your notebook. Be sure to highlight what it was that made you choose that variety over the others.

After you have filled in your crops on the Plant/Harvest Times worksheet you can add the days to maturity. That's the number of days it takes from either putting the seed in the ground, or putting the transplant in the ground, until you can expect the harvest to begin. You'll find the days to maturity with the variety name in the catalogs and on the seed packets you might pick up at the store. If it is something that is usually transplanted, such as tomatoes and peppers, the days to maturity will be from transplanting, not from when you started the seeds in a flat. If it is from transplanting, the cultural information for that crop should say so.

If you are transplanting, you need to know how many weeks ahead to start the seeds before setting the plants out. You'll find that in the cultural information from the seed company. Put that number in Column C. In Column D, put the date that you intend to plant in the garden. With a calendar handy, count back the weeks from Column C to determine the date to plant in the flat or coldframe for Column B. As I've mentioned, I start everything in coldframes these days. I find it much easier and the plants are naturally acclimated to the outdoor temperatures and amount of daylight by the time they are set out into the garden beds. Column E, the weeks in the ground until harvest, is easy since you already know the days to maturity. Divide the days to maturity by seven to find the number of weeks to put in Column E. Add that time to Column D, the date to plant in the garden, and you have the date you can begin to expect a harvest, Column F.

Length of Harvest

If you are new at growing, you probably have no idea how many weeks of harvest you will get from a crop. You might think that you plant every-thing at one time in your garden, and once things are producing you can pick it until frost. Well, it's a lot more interesting than that. Some things; such as garlic, potatoes, sweet potatoes, peanuts, and grains; will be harvested all at one time. Then the bed is ready for the next crop. Peppers will produce until frost. Frost doesn't bother the cabbage family—in fact those crops, as well as carrots and beets, taste better after they've been kissed by frost. To determine the number of weeks of harvest for Column G you will find hints in the seed catalogs, but ultimately you will learn from experience. Once you estimate how long the harvest will be, you can figure the end of harvest date for Column H. There is a column for notes on the worksheet. That's a good place to record where you got the seed from, what was so special about that variety, etc.

Learning about different varieties and how they behave in your gar-den makes you more connected with your land. Knowing the length of harvest ahead of time lets you know what to expect and allows you to see the whole picture for the year for your garden (and your diet). No matter what hints you've gotten from other sources, the learning is in the doing. Just get started and do it.

You'll find that peas and beans are available as bush and pole variet-ies. The bush varieties will give you a more concentrated harvest in a shorter time. I usually plan on a two week harvest for each planting of bush beans or peas. There is a difference of about ten days to maturity between the bush and pole varieties of the sugar snap peas that I grow. I've sometimes put a trellis down the middle of the bed, sown the pole varieties along the trellis and, at the same time, sown the bush variety along the edge of the bed. The bush variety would be ready to harvest first and, as those sugarsnaps are waning, the pole variety would be coming on.

When I was selling vegetables and wanted a continuous supply of green beans throughout the summer, I would plant bush beans every two weeks somewhere in the garden. Planting like that is called succession planting. As soon as one planting finished producing, the next planting was coming on strong. The bed with the previous planting could be cleaned up and planted to the next crop, which wasn't beans. One year, for our own use, I compared my regular bean variety, Provider, with one that had purple pods. The purple podded variety produced about as many beans as Provider; but it took three weeks, not two, for that production. It was clear that it wasn't done at the end of two weeks. These are the things you will find out from experience, your best teacher.

One summer, a few weeks into bean-picking, I attended a farm field day and met two men who had left corporate jobs to become farmers. This was their first year.

bush and pole beans

In our conversation I mentioned picking beans and the necessity to plant every two weeks, assuming they would be doing the same thing. Unfortunately, they hadn't grown beans in the years before they decided to sell them, and thought they just had to plant once. They had been trying to figure out why their beans, which started out so strong, weren't doing so well. They could plant again then, but by the time those new beans produced it would be the end of the summer.

Some crops are cut-and-come-again. Collards, kale, and Swiss chard are that way. They keep coming back after they're cut. In our hot, humid climate, it is hard to keep collards and kale through the summer, but I've had Swiss chard in the ground from one spring to the next by providing summer shade and winter protection. I use spring and fall plantings for collards and kale, harvesting the later planting through the winter.

Once they've gone through a winter, collards, kale, and chard will bolt, which means they will send up a seed stalk, flower, and produce seeds.

Many people manage lettuce as cut-and-come-again. That has its limits and lettuce will begin to get bitter, particularly if you have planted it in the spring and the weather has turned hot. When I grew lettuce to sell, I grew it through the summer with shade and plenty of water, but I only cut it one time. The bed that was harvested would be amended with compost and any other organic amendments that were needed and replanted with lettuce transplants. I might get three harvests from one bed throughout the season. Lettuce wouldn't be in that bed for the next two years. In order to have transplants each week, I would start seeds every week in an area set aside for seedling production. After three weeks they would be ready to be transplanted. I've had customers at the farmers market and chefs tell me that my lettuce was the best they'd ever had. I'm sure it was because it was always the first cutting. I let the lettuce plants grow out to large plants before cutting, making the most of every seed I planted.

cut and come again chard

Determinate and Indeterminate

When it comes to tomatoes, you should know if the varieties you have in mind are determinate or indeterminate. In my early years as a gardener I had to look that up each year, but now I remember that determinate varieties are determined to stop early and you don't know when the indeterminate ones will stop. The seed catalogs make that designation for tomatoes. I'm not sure if the tags on tomato plants in the garden centers indicate determinate or indeterminate, but if you are buying plants you can check the variety yourself in the catalogs or online. It can mean the difference between your plants producing for a month (determinate) or all the way to frost (indeterminate). I use determinate varieties for the tomatoes I intend to process so that I can have a lot at one time. Of course, I want to have tomatoes in the garden into the fall, so my garden also includes indeterminate varieties. Determinate varieties generally grow shorter and bushier than indeterminate varieties.

Plant and Harvest Schedule

My favorite worksheet is the Plant/Harvest Schedule. You will use the information from the Plant/Harvest Times worksheet to complete it. The numbers in the top row are the number of weeks before and after the last frost. Write the date of your last expected frost in the space below the zero. That column is for one week. Under the numbers in the blocks on either side of the zero, write the dates at two week intervals from that frost date. You could write in the dates of every week, but that gets crowded. Just know that each block in that row accounts for two weeks, except for the one with the frost date. Hint: write in the dates on one worksheet, and then get copies made. That way you only have to do it once. Since the first fall frost is important, find the week on your schedule that includes the first expected frost date in the fall and highlight it all the way down. Sometimes the timing for planting for fall and winter harvest is indicated as so many weeks before the first fall frost. With the first frost date highlighted, it is easy to determine recommended planting times.

Using the key I've provided, record when you will be planting and harvesting each crop. You will notice the faint gray lines dividing the columns into two weeks to help you stay organized. I've put "f1" and "f2" because some people plant the seeds in one flat and later transfer them to another before transplanting out. The "c" indicates that the seeds will go in the coldframe. The "p" indicates seeds planted in the bed and "t" indicates the transplant time. The "h" is for each week of harvest. There will be an h in each week that you expect a harvest. The "m" indicates that the cover crop, such as rye, was cut and left in the bed as mulch. There will be a two week delay before that bed can be transplanted into. If you don't like my key, make up your own. This is your plan. List all

Figure 7.2.
Plant/Harvest Schedule

PLANT / HARVEST SCHEDULE

Each block represents 2 weeks, except in the 0 column, which represents one week. To use this chart, write in the date of your average **LAST SPRING FROST** in the space below the 0. Then fill in to the left, the dates of each column at 2 week intervals BEFORE that date. Fill in the dates in the columns to the right of 0, the dates AFTER the last Spring frost. List the crops that you plan to grow and the beds they will occupy. Use the following key to indicate the weeks of your activities with those crops:

f1 = seeds into 1st flat   f2 = seeds into 2nd flat   c = seeds into cold frame   p = plant seeds into the bed   t = transplant into the bed   h = harvest   m = cut for mulch

Bed #	CROP	08	06	04	02	0	02	04	06	08	10	12	14	16	18	20	22	24	26	28	Notes

Homeplace Earth
Education and Design for a Sustainable World
www.HomeplaceEarth.com

Download this worksheet at http://tinyurl.com/mf4a33r

the crops in each bed, then list the next group, and so on. Having all the crops from one bed together makes a difference on this worksheet. You will readily see when the bed is empty and what goes into it next. You might not know what bed everything is in yet, but in Chapter 8 I'll talk about rotations and how to decide which crops go where on your garden map. For each crop, put the p's and h's, and whatever other letters you are using, in the spaces for the weeks that coincide with the times on the Plant/Harvest Times worksheet. I enjoy color, so I color in the rows, with a different color for the time in the flat or coldframe, growing out, and harvesting. Usually it's yellow for the flat/coldframe time, green while it's growing in the bed, blue for the weeks of harvest, and brown for the weeks of delay when the mulch has been cut until the next crop goes in, but that's just me. Make this plan your own.

Once this worksheet is complete, you can look at it and know what and when you should be planting and what you should expect to be harvesting each week. Granted, not all 52 weeks fit here, but it will suffice for most folks. The last crop shown in the fall will be either a cover crop or garlic and onions, which will grow through the winter; or crops such as collards, kale, carrots, and beets; which will be harvested through the winter. By that time, the planting is over. If you are planting through the winter, you can use this idea to make a worksheet specific to your needs.

Now that all your crops are shown, you might notice a time when there is not much to harvest. You could rework your plan to make sure you have food when you want it. Often it is just a matter of adjusting your planting times.

Plan for Special Events

If you know that you will be away on vacation, you could highlight that week. It is really disappointing to watch your vegetables growing and then miss the optimal harvest time. If your green beans, or other such crop, show up with harvest at a time when you will be away, adjust your planting time. One of the best things about staple crops, such as the potatoes and grains, is that you have some leeway as to when to harvest. You do have to pay attention, and more than a week's delay might pose a problem, but it's not like green beans, cucumbers, and zucchini, which quickly grow past their prime.

On the other hand, you might want to highlight a week when you will be home and planning a party. What better way to celebrate than with food from the garden? It could be your birthday or some other occasion. When our daughter, Betsy, got married in 2010, I grew some of the food and used this worksheet to make sure we had the timing right. It was a big help. Sometimes I use this to work out a schedule for just a few crops. It has any number of uses. It is invaluable to use if you are growing for the markets. Highlight when the market season begins and ends and make sure you have something ready to harvest each week.

This is your *proposed* plan and, no doubt, you will want to know how it all works out. To know your *actual* plant and harvest times, you could record right on your original schedule when all these actions occurred, or make another schedule with all your crops and record it there.

8

Rotations and Sample Garden Maps

Now that you have an idea when things are going to be in the garden beds, it might be a good time to fill in the crops on your garden map. It would be easy to just decide where your crops go according to height—short ones on the south side and tall ones in the back. I've seen some plans like that and I am always left wondering—but what about next year? As far as I'm concerned, garden maps need to show rotation arrows. These are arrows that show where the crops in one bed move to the next year. A garden growing a sustainable diet will have something growing in each bed all twelve months of the year. It is particularly important to make sure the cover crop that is planted in the fall is the crop you want in that bed in the spring.

Height is still a consideration, but when the sun is high in the sky during the summer months, it may not be as much of a problem as you might think. Observe how much of a shadow your tall crops cast at different times during the growing year. My solar food dryers are on the northwest side of a large maple tree. They get plenty of sun until September when the shadow from that tree extends to the dryers. That's when I move them. If a sun-loving crop were in that spot it would be fine, as long as it was harvested by early September.

You will want a tight rotation. By tight rotation, I mean that when one crop is harvested, the bed is amended with compost and anything else it needs, and the next crop is planted without delay. The easiest way to accomplish that is to have it all decided ahead of time and shown on your garden map. You will know what is planted where and when, and would have bought the seeds well in advance. I will be talking about seeds in Chapter 9.

In deciding where everything goes, it is important to not plant crops in the same place they were the year before. Different crops require different things from the soil. If you planted everything in the same place every year, the soil would become depleted of certain nutrients. You can fertilize, but moving things around is a more balanced approach. That is, in addition to having your soil tested and adding any organic amendments, of course. Also, if you plant things in the same place each year, pests and diseases will accumulate. The pests know just where their favorite crops are and use those spots to raise their young, with the assurance that food will be there when they need it. There is no break in the disease cycle, either. You can fix that by rotating your crops—not planting them in the same places each year. In fact, you want to avoid planting anything in the same family there for the next couple years. With that said, I do have to mention that many sources recommend that tomatoes can stay put year after year. Whoever recommends that must not live in Virginia. With our hot humid summers, there are many tomato diseases here and I rotate my tomatoes each year. I've provided a list of some common crops and their kin, so you know which ones are related.

There are more things to know than simply not planting the same crop family in a bed. Legumes—peas and beans—leave behind some nitrogen for the next crop, so they might be followed by crops that are a little hungrier for nitrogen than others. If you have mulched a crop, the mulch will have discouraged weeds, leaving a cleaner seedbed for the next crop. It might be advantageous to plant something like carrots in that bed next.

In *The New Self-Sufficient Gardener*, Seymour divides his garden into four sections with groups of crops divided as (1) Miscellaneous,

Figure 8.1. Crop Families

Alliums
garlic, leeks, onions

Beet Family
beets, spinach, Swiss chard

Cabbage Family
broccoli, Brussels sprouts, cabbage, cauliflower, collards, kale, kohlrabi, mustard, radish, rutabaga, turnips

Carrot Family
carrots, celery, parsley, parsnips

Cucurbits
cucumbers, gourds, melons, summer squash, winter squash

Legumes
peas, beans

Morning Glory
sweet potatoes

Nightshades
eggplants, peppers, potatoes, tomatoes

(2) Roots, (3) Potatoes, and (4) Peas, Beans, and Brassicas. The crops in each section rotate to the next spot each year. In *New Organic Grower*, Eliot Coleman has an excellent chapter explaining rotations with an eight year plan. His plan inspired Pam Dawling to make a ten year plan and put it on a pinwheel. You can read about that in *Sustainable Market Farming*. If you are like me, you never have the one perfect plan. You work out rotations and that plan may do you well for a few years; then you decide to grow more or less of something or throw some new crops into the mix. If you understand how it all works, it will be easier making adjustments.

In my garden I had grown peanuts after wheat and rye for years. The grain harvest finished just in time to get the peanuts in. I thought that, since peanuts were legumes, it didn't really matter what preceded them. I started to stay away from following grains with peanuts because the grain beds attracted voles, which are my biggest garden pest. One year, when I planted peanuts in a bed after garlic and onions, the half that was planted after the onions did considerably better than the half that followed the garlic. At first I thought I had planted two different varieties, but I hadn't. I checked my garden records and realized that the onions, which were planted from sets in the spring, had followed Austrian winter peas. The garlic was in there since the fall. I adjusted my garden plan, and that fall I planted Austrian winter peas in a bed where peanuts would go the following year. My yield that next year was almost three times in that bed compared to the peanut crop following a bed of kale, onions, and garlic. When you are in your garden, make it a practice to notice little things like that. Write it down and find out what happened. It only took a glance at my garden map to realize the difference in the onion half of the bed. I could immediately see that there had been Austrian winter peas before the onions. Then I vaguely remembered reading about a legume cover crop before peanuts, which is also a legume, so I checked *How to Grow Vegetables & Fruits by the Organic Method*, my favorite reference book. That book actually suggests two soil building crops before peanuts. Okay, so there goes the guideline about not following with the same crop family. I had avoided legumes before peanuts for just that reason. Now I know better. Become a student of your own garden and let it teach you.

Read all you can and make a list of guidelines you come across. You might keep that list in your garden notebook so you'll know where it is. At the same time, begin a list of guidelines drawn from your own experience in your garden. The companion planting charts list potatoes and cabbage as friends. Since the voles cause problems for me, I thought planting potatoes with cabbage might offer some protection. Not a chance. The voles took out the potatoes every time I tried that. I had also tried oilseed radish as a cover crop before potatoes, thinking it would offer an early fertile seedbed for the potatoes in the spring. That's when it really hit me how much the voles like to hang around the cabbage family plants. There were vole holes galore, just waiting for my potatoes. One item on my personal list of guidelines is to keep the potatoes away from the cabbage family. If you don't have any problems with voles, potatoes and cabbage might just work well for you.

I map out my beds with arrows to show how they will rotate. The examples of garden maps with crops, complete with the times they will be in the beds, are here to get you started thinking about how you can plan your garden. I will explain my reasoning behind the choices, how I would manage the cover crops, and give suggestions for variations on what is presented. When trying to decide how much space to give everything, it helps to map it out on graph paper. You can find suggested spacing in the seed catalogs, on the seed packets you buy in the store, and in *How To Grow More Vegetables*. The more intensively you plant, the more important it is to make sure your beds have enough fertility and that you provide enough water if rainfall is lacking. The dates on these maps are based on the frost dates in Zone 7—last frost April 25 and first frost October 15. The Plant/Harvest Schedule (Figure 7.2) will help you determine the times to plant in your area.

Transition Garden

The three bed garden in Figure 8.2 is called the Transition Garden because it shows a garden with crops most people are already comfortable with, plus some new ones and cover crops. It combines what you already know with growing staple crops. The spacing of the main summer crops

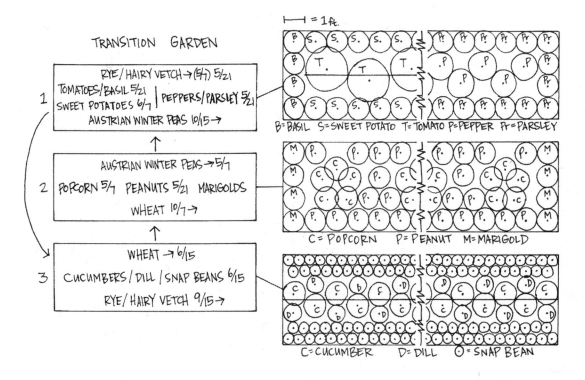

⊢—⊣ = 1 ft.

TRANSITION GARDEN

1 | RYE / HAIRY VETCH → (5/7) 5/21
TOMATOES / BASIL 5/21 | PEPPERS / PARSLEY 5/21
SWEET POTATOES 6/7
AUSTRIAN WINTER PEAS 10/15 →

B = BASIL S = SWEET POTATO T = TOMATO P = PEPPER Pr = PARSLEY

2 | AUSTRIAN WINTER PEAS → 5/7
POPCORN 5/7 PEANUTS 5/21 MARIGOLDS
WHEAT 10/7 →

C = POPCORN P = PEANUT M = MARIGOLD

3 | WHEAT → 6/15
CUCUMBERS / DILL / SNAP BEANS 6/15
RYE / HAIRY VETCH 9/15 →

C = CUCUMBER D = DILL ◯ = SNAP BEAN

is shown in detail. What you see is the beds with what they contain for the calendar year. The rye and hairy vetch cover crop in Bed 1 was planted the previous fall when the combination of crops that have now rotated to Bed 3 were in Bed 1, leaving the rye and hairy vetch to over-winter there. In this plan the rye/vetch cover crop is cut and left to lie on the ground as mulch on about May 7, or when the rye is shedding pollen. The vetch would be flowering by then. Vetch can get pretty rangy and tangle in the rye. I only use these crops in combination if I'm going to be cutting them early like this. If I was to leave the rye go to seed, I would have planted Austrian winter peas as the legume and pulled out the winter pea plants for compost material when they flowered, at least a month before the time to harvest the rye seed. In this case, it doesn't matter. I like vetch before tomatoes and sometimes plant only vetch as the cover crop preceding tomatoes. The legumes don't have the heavy root system that the grains do, so when they are cut, the bed is friable and ready for planting the next crop. I always leave the roots right there

**Figure 8.2.
Transition Garden**

in the soil. With the grains cut at pollen shed like this, you need to wait two weeks for the roots to start decomposing before you can transplant into the bed. Even then, it would be transplants only. The grain bed, cut at this time, is not ready for seeds even two weeks later. If hairy vetch was the only cover crop, I could transplant the tomatoes the same day I cut the vetch, leaving the vetch as mulch. The legumes don't have as much carbon, thus don't have the staying power as a mulch that the grain crops do. You see the date I cut the rye/vetch as (5/7) because it is not ready for the next crop yet. May 21 is the date the bed is ready for transplanting.

On May 21, tomatoes, peppers, basil, and parsley will be transplanted. Tomatoes and sweet potatoes are in one half of the bed, with peppers and parsley in the other half. The sweet potatoes could go in the same day, but since I wait until the last week of April to start sweet potatoes in the coldframe, I don't have them ready for the garden until June 7. The rye/vetch mulch is providing a good ground cover until then. Sweet potatoes like hot weather, so don't be in too much of a hurry to put them in. The tomatoes would be on a trellis in the middle of, in this case, a four-foot wide bed, with sweet potatoes planted along the outer one foot on each side. I use a section of fence for a tomato trellis. Peppers are planted with a border of parsley. Most people like to have tomatoes in their garden and basil is their natural companion, both in the garden and on the table. Make sure to plant basil with your tomatoes. You can eat tomatoes fresh, and they are easily preserved for later. The same goes for the peppers and parsley. Peppers, parsley, and basil can be dried. I only have the freezer space that comes with my refrigerator, so I don't generally freeze vegetables, but I do chop some peppers for the freezer. They need no other preparation. If you are canning spaghetti sauce, or making it from dried ingredients, basil, peppers, and parsley make great additions to the tomatoes.

The sweet potato vines will provide a living mulch for the tomato plants once the rye/vetch mulch disappears back into the earth. Before the first fall frost is expected, dig the sweet potatoes and put the vines in the compost pile. If you wait until everything is killed by frost, the vines will not have as much biomass for compost and you will be losing some of the goodness.

The tomatoes in this example are an indeterminate variety, hopefully producing all the way through the season. If you plant a determinate variety, all the tomatoes will be produced within a smaller window of time. If you have a shorter growing season than I do—your last spring frost is later and first fall frost is earlier—you might need to plant a determinate variety to get the full harvest in. With a shorter season, you could go with just the hairy vetch cover crop so you don't have to wait the two weeks before planting. Look for varieties of the other summer crops with early maturing times if you have a shorter season. Austrian winter peas are the cover crop to be planted in this bed when the summer crops are finished. That is the legume that can be planted the latest in the fall and still produce well. Other types of field peas are not winter hardy in my climate, but might be in yours. Rye is the carbon-producing crop that can be planted the latest in the fall.

The crops from Bed 1 were planted in Bed 2 last year, leaving Austrian winter peas there at the end of the year. This crop will put on only a little growth in the fall. You can cut the ends of the plants—the pea shoots—for stir fry, or to eat in your salads through the winter. In early spring, the rate of growth of the plants will speed up. According to this plan, these winter peas were planted in mid-October. If they were planted around September 1, they would have grown a lot and maybe even flowered before the weather turned cold. In that case, they might have winterkilled. I like to let winter peas grow until at least April 1, but if they can go longer, all the better. In this case, in Bed 2, they are in until the first week of May, about the time they will be flowering. The pea vines can go right to the compost pile. It would be good if you have some carbon material to add at the same time. If you grew Jerusalem artichokes and dug them through the winter, you could use those stalks as your carbon material. Otherwise, you might have saved some stalks from corn or sunflowers from the previous fall to add now.

I chose popcorn for this example because it is fun to grow and eat. If you have only grown sweet corn, growing corn out to the dry stage on the stalk is a new experience. The popcorn, like the corn that you would grow to grind for cornmeal, will stay on the vine until the husks are dry and the plant looks dead. When you cut the stalks at harvest you can use them as a fall decoration before they become compost material. I use a machete to cut the cornstalks, first to cut them out of the bed, then to cut them in lengths convenient for compost material. We occasionally get summer storms that can knock the cornstalks down if they are planted intensively. Sometimes they stand back up by themselves, and sometimes not. In any case, they are never the same. To avoid that, I plant corn in circles, as you see in the planting diagram. Corn will germinate pretty quickly, as will beans. Crows love to eat newly germinated corn seedlings and I found that if I transplant it rather than put the seeds in the ground, the crows are not a problem. Generally, I put five corn transplants in an 18" circle, with the center of each circle four feet from the next. That leaves room for the wind to blow between the corn circles without taking down the corn in those summer storms.

Planting the corn in circles leaves half the bed available to something else. In this case that other crop is peanuts. The peanuts are transplanted in the rest of the bed on 12" centers. Marigolds are at the ends of the bed, both for pretty and because I think they help the peanuts. Just as with the sweet potatoes, I want to let the peanuts grow until there is danger of frost, but not until after the frost hits. The peanuts themselves will be growing underground. I dig the whole plant, shake off the dirt (the peanuts will stay attached), and hang bunches of the plants in the barn for at least a few weeks to let the peanuts dry. The plants will dry, becoming peanut hay. The peanuts can then be picked off to eat. Alternatively, you could pick off the peanuts at harvest and use the plants as green material in the compost, along with cornstalks as the carbon material. The nuts will need more drying and you can accomplish that by laying them out on screens. If you do that, make sure it is in a mouse-proof area. The mice will steal them, but to you it will look like they just disappeared.

You can leave the corn roots in the bed, amend with compost and anything else that is needed (as you would before every new crop) and broadcast the wheat. I use a four-tined cultivator to chop the seeds into the soil. You could also use a hard rake. If the weather has been dry and is not promising rain, it might be better to make furrows for the seed, cover them with soil, and water well. In fact, at times when we are really in a dry pattern, I've watered the furrows before I covered them to make sure the seeds got off to a good start. That tip goes for any crop, not just wheat.

That brings us to Bed 3 and the wheat that was planted there the previous fall. It can be cut with a sickle when the grain is mature, which is about mid-June at my place. Watch it closely as it approaches that stage, so you can see what is happening. Pull some off the seed heads and taste it. You could buy some wheat berries at the health food store if you need something to compare it to. When the time is right, cut the stalks close to the ground. I talked about harvesting grains in Chapter 5. Once the wheat is harvested, the bed is clear except for the stubble. The plants have reached the end of their lives and that stubble is not firmly attached to the roots, which are already on their way back to the earth. The next crop can go directly in the ground, with the stubble still there, as long as the bed has been cleared of any weeds that may have crept in. In this case, the next crop is dill, cucumbers, and snap beans.

The cucumbers are grown along a piece of fence, or some other trellis, installed down the middle of the bed to save space, and snap beans are planted on the sides. The June 15 planting date indicates that they will be planted when the wheat is harvested. You need to have plenty of foliage on the cucumber plants to plant this way. If there is not enough leaf cover, the cucumbers will be scalded by the sun. In that case, you would have to forgo the beans and plant the cucumbers without the trellis, letting them sprawl in the bed. Dill is beneficial to the cucumbers and many other plants. It is also an ingredient in pickles, which you would want to make. You can find directions for Sour Pickles in *Wild Fermentation* by Sandor Katz. These are the dill pickles that you would most associate with a deli. I like to make a gallon jar (or larger) of these at a time in the summer and have it on the kitchen counter to eat from as we want them. The snap beans can be eaten fresh or canned. Rather than canning them, to save fossil fuel you can salt them down in a crock as Seymour shows in *The New Self-Sufficient Gardener.*[1]

By September the cucumbers, beans, and dill will be ready to give way to rye and hairy vetch, which will be broadcast and chopped in, like the wheat seeds. Next year, tomatoes, sweet potatoes, peppers, and parsley will be in this bed. Remember, I'm just giving suggestions here, but if you grew these crops you would have tomatoes, peppers, and parsley to put in your solar dryer, which I'll be talking about in Chapter 11. You can just hang the basil to dry in your kitchen. You'll learn fermentation techniques with the cucumbers and beans, and learn to grow corn, wheat, and peanuts to their dried state. You could grow a variety of bush beans that is suitable to use as a dry bean and let the beans grow out to the dry seed stage, rather than picking them as snap beans. The rye can be planted in October if the beans occupy the bed until then. You could use either hairy vetch or Austrian winter peas as the companion legume with the rye. If you are using vetch, as shown on the garden map, you would want to plant the rye/vetch by early October. If the planting extends later than that, make Austrian winter peas the companion legume with the rye. I mentioned in Chapter 5 that it is beneficial to plant a legume with a small grain, but in small amounts, so the legume doesn't overtake the grain. In this plan, the crop will be cut down and left as mulch for the tomatoes, so the amount of legume isn't so much of a concern.

Managing a garden this way is different than managing a garden with a tiller. The ecosystem benefits from the diversity of crops and the fact that the ground isn't churned up on a regular basis. There is some learning to do on your part, for sure, to become comfortable with this system. Once you are, you will find that you are gardening smarter, not harder, which is always a good thing.

Quartet of Beds

I've added some things in a four bed plan (Figure 8.3) that go beyond basic gardening skills. In Bed 1, you will see oats that are there to winter-kill ahead of the potatoes. In order for that to have happened, the oats must have been planted between mid-August and mid-September to allow enough time to put on good growth and maximum biomass before the cold weather stops it. It will look great until January. A guideline for you to follow would be to plant the oats at least a month or more before your first expected fall frost. If your winter is unusually mild, or for some reason this bed is in a protected place, the oat plants might not die as planned. Potatoes are the next crop intended for that space. I would plant oats in early spring if I wanted them for grain. Oats have a hull on them that would be hard to remove with the threshing methods I use for rye and wheat. If I was growing them to eat I would grow a hulless variety.

My compost piles are on a garden bed and part of the rotation. One year I planted oats, intending them to winterkill, on the south side of the compost bed. The compost piles were enough protection that the oat crop survived and thrived through the winter. It would have been great to have had a bed of winter greens in that spot, to eat through the winter, but not the oats that I wanted to winterkill.

Instead of oats in Bed 1 of the Quartet of Beds, I could have planted radish (oilseed, fodder, or Daikon) to winterkill, but remember I said they attracted the voles and I didn't want that with the potatoes. Planted at the same time I recommended for oats, the radishes can get quite big and you can harvest some to eat—until January. They will die when the really cold weather moves in, leaving lots of decomposed biomass, holes for air and water to come in, and a clean seedbed. With these winterkilled crops, the bed is ready at the beginning of March. Things like onions, peas, lettuce, spinach, collards, and kale could go there, but in this case potatoes—that great calorie producing crop—are going in. Make sure the weeds don't move in before you get the potatoes planted. The dried oat biomass (foliage) will be lying on the bed. Providing it was a thick stand of oats and you didn't wait too long to plant your potatoes, you can just pull the dried biomass away and put the potato pieces in the ground using a trowel. The roots of the oats would have decomposed by that

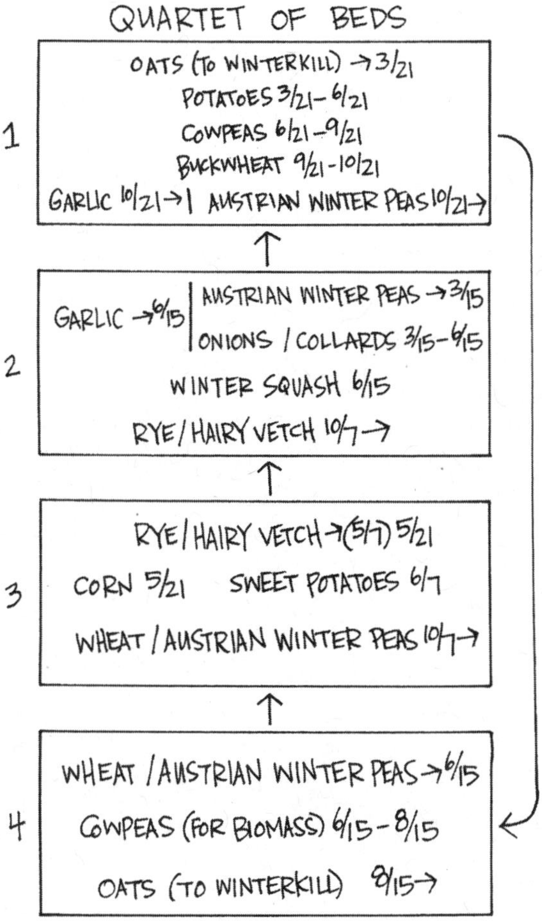

Figure 8.3. Quartet of Beds

time. The oat biomass can be put back on top of the bed as a light mulch and the potatoes will grow right through it. If the oat planting was not a thick one, as soon as the soil begins to warm, spring weeds will come up that will need to be dealt with before the potatoes go in. Potatoes are dug all at one time when the plants begin to die back.

Cowpeas produce well when planted after potatoes and they are next in the rotation. The date, mid-June, is the same for taking the potatoes out and for planting the cowpeas. To insure a tight rotation, the next crop goes in without delay. There will be a gap of about a month from when the cowpeas come out to when the garlic and winter peas are planted. Buckwheat is a great crop to plant to fill that spot. It is a summer cover crop that will be killed with the first frost. It doesn't produce much biomass, but it keeps the weeds out and provides nectar for the bees on the flowers that appear within 30 days. Keep some buckwheat handy for any place that will be empty for about a month. It can be easily pulled out when the next crop goes in and it leaves the soil in great condition.

The time to plant garlic in Zone 7 is between October 15 and November 1. Austrian winter peas are planted on the two thirds of the bed that are reserved for onions and collards for the following spring. If the cowpeas were out in time, oats or radish could go in that spot instead, with buckwheat only in the part of the bed that will be occupied by the garlic.

In Bed 2 you can see that the garlic planted the previous year is there and will be harvested in June. The winter peas gave way to onions and collards in March. This gives you garlic and onions to eat and store, and calcium-rich collards to eat all spring. The onions planted here could have gone in as sets (small onion bulbs for planting) or onion plants. Winter squash is next. Find a variety that is best suited for your area. I grow butternut because it does well in my hot, humid climate, fends off the squash bugs better than other varieties and, properly stored, can last all winter with little attention. In the fall, rye and vetch are broadcast in the bed.

As before, the rye/vetch cover crop is cut as mulch for the corn. Instead of peanuts, as in the Transition Garden, Bed 3 has sweet potatoes planted two weeks after the corn. I like to let the corn get off to a good start first. The sweet potato vines will grow to cover all the space not taken up by the corn circles. If you don't have trouble with your corn being damaged by summer storms, you might want to plant corn intensively, filling the whole bed without a companion crop. I enjoy planting things together and have had good luck with sweet potatoes under corn. Wheat and Austrian winter peas follow the corn and sweet potatoes.

In Bed 4, cowpeas are planted after the wheat harvest. In this case, however, they are grown for a biomass crop for the compost pile. They will be cut at about sixty days, or when they are flowering, and the biomass will go to the compost. Wheat straw can be added to the compost pile at the same time.

Garden of Ideas

If you are serious about growing compost crops, in addition to growing your food, the Garden of Ideas map (Figure 8.4) is for you. There is a lot here to show you how to make your garden and diet more sustainable. You've just seen the crops in Bed 1 on the previous map. In Bed 2 in the Garden of Ideas, in early March, red clover is broadcast into the wheat that was planted the previous fall. Red clover will grow under the wheat and not be a problem with the wheat harvest. When the wheat is cut, the clover will continue to grow and you should get one cutting off of it the first summer. As the map shows, it stays in the bed through two cuttings the following year. I don't have dates there, because it all

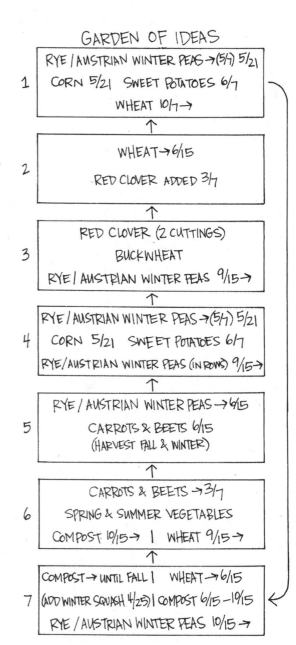

Figure 8.4. Garden of Ideas

depends on how you manage it. If it dies out and the bed is open, you could throw some buckwheat in until time to plant the rye.

Bed 4 shows corn and sweet potatoes again, as you saw in the Quartet of Beds, but this time the following cover crop, which could be wheat or rye with winter peas, is planted in rows, not broadcast. When that grain is harvested, the stubble will be in rows. You can lightly hoe a furrow for seeds between each set of stubble rows and plant carrots and beets there, leaving the stubble in place. Water well and keep a look out for the seeds to germinate. If you have spots with poor germination, plant again. Soon everything will be up and growing and you will have carrots and beets to eat through the winter. I don't add a cover over the bed because I don't want the voles to move in. If our winter was a little more severe, I would put leaves or straw over the bed once the weather turned cold. I would want the voles to find other winter homes before I do that. If we had more severe winter weather I would put a low plastic tunnel over that bed as I do over the collards and kale I grow. The carrots and beets will be harvested by early March. If there are any left, you will need to take them out before they send up seed stalks, unless you want to let some go to seed. Vegetable crops can go in this bed from March through the summer. In the fall, compost is built on part of the bed as the other beds are cleaned up, and cover crops are planted. The next year winter squash will be planted around the base of this pile, covering it with vines and shading out any weeds. This pile will be ready to use next fall. Any compost made in Bed 7 that is not ready to use yet gets moved to Bed 6 at this time. This is the compost that was started in the summer. It will be ready to use early in the season next year. If you don't need to use the whole bed for compost, you can plant wheat or rye in the empty space in Bed 6. That space will be needed for a compost pile the next year after the grain harvest.

In Bed 7 you see the compost area for the current year. Winter squash is planted around the base of the pile built the previous fall. The squash vines grow over the pile, shading it and discouraging weeds. In the fall, the finished compost (under the winter squash vines) will be used on the garden, the summer-built compost will be moved to Bed 6 (the only time it is turned), and rye and winter peas are planted. I have my best corn yield in the bed that follows the compost. Until you really put this system to use, it is a balancing act to have carbon and nitrogen biomass from your garden in the proportions you need. Keep with it and you'll get the hang of it. When you do, it will be transforming. In my garden in an area with eighteen beds, one is for compost in the rotation, exactly as you see here. In that rotation of eighteen beds, I have one planting of red clover each year (plus the bed where it overwintered). Actually, it's half red clover and half alfalfa because I'm studying both. I have packed a lot of ideas in this map so I could show you how compost piles, second year biomass crops, and winter-harvested carrots and beets work in rotation. You wouldn't necessarily be planting a group of seven beds just like this. These are ideas to incorporate into your larger garden. You can see how all this works in my video *Cover Crops and Compost Crops IN Your Garden.*

9

Seeds

A SUSTAINABLE DIET begins with sustainable seeds. That means seeds grown and saved in a manner that is good for the earth and that will perpetuate their genetic heritage. If you are not into saving seeds yourself yet, you need to find them from a source that follows those guidelines. You will be looking to seed companies that have signed the Safe Seed Pledge to "not knowingly buy or sell genetically engineered seeds or plants." You will also be looking for open-pollinated seeds—seeds that are bred from parents of the same variety. When you plant them, the offspring are the same as what you planted. Seeds in the catalogs are open-pollinated or hybrid. Hybrid seeds are bred from parents of different varieties. Seed companies do that to produce a variety with distinct characteristics, often marketed to a large geographic area. The hybrid varieties will be designated with an F1 near the name or will just have "hybrid" in the listing. The offspring of the plants produced from those seeds will not necessarily have the same characteristics as the parent. You will have to go back to the seed company each year for hybrid seeds. You could de-hybridize a variety yourself, with careful selection, but it would take about seven years to have a stable variety. Heirloom varieties are open-pollinated.

What you are looking for is not a variety that a seed company thought would be marketable. You want seeds that will produce well in your area, specifically in your garden, no matter what the weather conditions. If you have a drought year or abundant rainfall, these are the plants that will not fail you. If you are saving them yourself and you save seeds from the plants that thrive at your place, you are one step closer to sustainability. If you are not saving them, you will need to look for a seed company in your region that can supply you with seed from their own gardens or from local/regional growers. That company will even tell you about their growers in their catalog. When you go to conferences you might even meet these people—both from the seed company and the growers. Here in Virginia we are fortunate to have Southern Exposure Seed Exchange. Many of their varieties will do well in other areas and all are well adapted to the mid-Atlantic area, but some are designated to be especially well-suited to the Southeast and the hot, humid summers you find there. Take some time to find sources of seeds suited to your climate and region.

If you are buying plants to get started, consider buying the ones grown in your area. Farmers markets are often good sources, but you will need to either put your order in early, or buy the plants when you see them at the market. There are no guarantees that they will be there the next week, unless you talk to the farmer about saving some out for you.

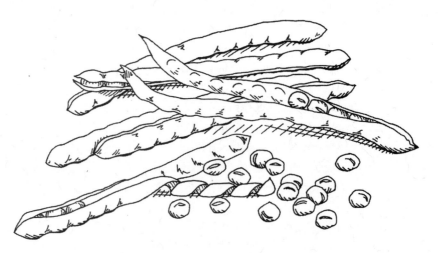

If you find that there are not enough sources for locally grown plants, you may have discovered a niche you could fill in the future. Often the plants for sale, even in small businesses, are trucked in from a distance. That means that in a large region all the plants could come from the same place. Not so long ago, there was a disease problem with tomatoes that raged up and down the East Coast. It was spread by tomato plants from just such a source. Even though you might notice your local farm store selling these types of transplants, talk to them and ask if they also have sources from local growers. Maybe their sweet potato slips come from just down the road. If not, they might direct you to someone who can answer your questions or supply you with plants. If enough people ask these kinds of questions, the store management will see that there is a need that is begging to be met. If they think you are a little crazy for being concerned about such things, that's okay—make friends with them anyway. One day someone else will come in with the same concerns and they will refer them to you. Before you know it, that store will begin to stock the things that you and your new friends have been asking for.

Seed Inventory

Before you order more seeds you need to know what you already have. To find out, take an inventory. I used to just write it all down on a piece of notebook paper each year, but once I got more organized I made the form that you will find here (Figure 9.1). The more information you put on the form, the more it will help you through the season. I keep my inventory in my garden notebook and find it helpful to refer to if I'm wondering about varieties I have on hand, what I've ordered, etc. Once you order new seeds remember to add them to your inventory. I frequent events that have seed exchanges. If those new seeds are added to my inventory, it is more likely that I'll remember to use them.

In your inventory, besides listing the crop and variety, determine the amount of seeds you have. Sometimes I put the weight there, sometimes the count, and sometimes I just put "lots", "enough" or "not enough". Knowing the source of your seeds is important, along with the year it was grown or sold. If I've bought the seeds, I put the initials of the seed

SEED INVENTORY FOR _____

(year)

Figure 9.1.
Seed Inventory

Crop	Variety	Amount	Source		don't buy	DO buy	Source	Amount	$

Homeplace Earth
Education and Design for a Sustainable World
www.HomeplaceEarth.com

Download this worksheet at http://tinyurl.com/mf4a33r

company and the year I received them, such as "SESE '13". If I grew them myself, I put a star and the year they were grown—"* '13". Seeds lose their vitality as they get older, so don't keep them forever. If you have a stash of seeds that have some age on them, do a germination test to see if they are still viable. If not, throw them to the wind, or at least in your backyard, chicken yard, or birdfeeder. Even if they don't grow, they can be picked up by birds or composted back to the earth.

On the Seed Inventory I left a column blank. If I've done a germination test, I put the germination rate in that column. Otherwise, I might put the days to maturity there for that variety. I can make my seed shopping list as I do my inventory. With the columns for "don't buy" and "DO buy", I know just what I need. As I list each crop, I leave an extra line or two open if I know I'm going to be trying some new varieties not shown. Sometimes I look up those new varieties while I'm thinking about it as I'm doing the inventory, and sometimes I sit down with the catalogs later. Either way, there is space to put the source, amount of seeds to buy (packet, pound, etc.), and how much it will cost for anything I might need to order.

How Many Seeds?

The germination rate, how much space will be planted, and how many plants will go there, are all considerations in deciding the amount of seed you need. You can see from my worksheet, Seeds and Plants Needed (Figure 9.2), that you can easily make those calculations. On this worksheet, list all the crops that are planted. Use your garden map for that. Some crops are shown on your map in places where they are only harvested, having been planted the previous year. Only count them when they are planted. You might have multiple beds of the same crop. Combine the bed area on this worksheet so you know the total area that you need seed for.

Most likely, not all the seeds will germinate, so you need to determine an estimate to work with. Sometimes the germination rate is listed on the seed packet, along with the month and year the test was done. If you haven't done a germination test yourself, and the rate is not listed

_____
(year)

Bed #	Crop, Variety	A	B	C	D	E	F	G	H	I	J
		% germination	# of seeds per ounce	centers in inches	sq. ft. needed per plant	area to plant in sq. ft.	# of plants needed	# of plants with insurance	# of seeds needed	weight of seeds needed in ounces	Source or Notes
		(expressed as a decimal, i.e. 95% = 0.95)			(C x C) / 144		(E / D) x 1.13 for offset spacing	F plus 20% (F x 1.2)	(G / A)	(H / B)	

Figure 9.2.
Seeds and Plants Needed

Homeplace Earth
Education and Design for a Sustainable World
www.HomeplaceEarth.com

Download this worksheet at
http://tinyurl.com/mf4a33r

on the seed packet, you could go with the established legal minimum germination rate, which you can find in the Master Charts in *How to Grow More Vegetables*. Put the germination rate you are using, expressed as a decimal, in Column A. The Master Charts are a great reference for this worksheet, since you will also find the number of seeds per ounce and suggested spacing for each crop, which are needed for Columns B and C. The spacing, "centers in inches", is for planting in a grid pattern, with everything equally spaced. If you are planting in a different pattern, use your own figures.

Using equidistant (also called offset) spacing, the formula in Column D helps you figure the space each plant needs. In that formula the square inches needed (C × C) is divided by 144, the number of square inches in a foot. If you have already put your crops on your garden map, you know how many square feet are needed for each crop and you can put that in Column E. If the plants were all lined up in straight rows, like so many boxes, you could determine the number of plants needed for that area from the number in Column D. However, with equidistant spacing, the plants are nestled together. You can see that in the expanded planting diagrams in Figure 8.2. Planting that way, you could get up to about 13 percent more plants in the same space. Divide E (area to plant) by D (square feet needed per plant) and multiply that by 1.13 to find the maximum number of plants you can fit in that area with equidistant spacing. Put that number in column F.

Not every plant will be an excellent specimen, and you want to put out the best, so allow for some extras. In this case 20 percent extra is allowed. Multiply F (number of plants needed) by 1.2 to find the number of plants needed with 20 percent for insurance. That means that for every ten tomato plants that have germinated and grown, you will have two extra so you can pick and choose the best. Save the extras until you know that things are going as planned in the garden after you have set those plants out. It could be that something comes along and eats them in the first week. In that case, you would need those extras. Suddenly, second best becomes the plant of choice.

To determine the number of seeds needed for that number of plants (column H), divide G (the number of plants you need with insurance) by A (the germination rate expressed as a decimal). The weight of seeds is the number of seeds needed (Column H) divided by the number of seeds per ounce (Column B). The seed catalogs will indicate how many seeds are in their packets for each crop. Sometimes you will see the seed weight listed as grams. There are twenty-eight grams in one ounce. There is always something you'll want to remember about this crop or these seeds. You can put that information in Column J.

You can probably find online calculators that will supply you with the amount of seeds you need, but working these calculations yourself gives you an understanding you wouldn't have with a computer generated list.

Develop reference material specific to your needs. I have a chart for my own use that lists all the crops I might grow in my garden with columns for the spacing I use, seed needed per 100 ft², the yield figures from *How to Grow More Vegetables*, the yield I've obtained in my garden, and the friends and foes of each crop. That chart is in my garden notebook and serves as a quick reference. A chart with the information you tend to look up each year, specific to your garden, is a great resource.

Germination Test

If you want to have a more accurate number for the germination rate, you can test your seeds yourself. It's a good thing to do if you have older seeds or seeds that have been stored in less than desirable conditions. Many times it is suggested to use a paper towel to put the seeds on. We don't buy paper towels and I don't like them for germination tests anyway. I use coffee filters for germination tests and find they hold up much better when wet. (I don't normally buy coffee filters, either, but I have an inexpensive package of filters I bought years ago for seed germination.) Write the date, name, and origin of the seeds on the filter with a pen while it is still dry. Wet the filter and wring out excess water. Put at least ten seeds (make sure you know how many are there) on the filter. Fold it up and put it in a container with a lid—a wide-mouth jar will do. You can do many tests at the same time with each variety of seed on its own filter in the jar. Everything needs to be moist so the seeds will germinate; if necessary, add a few more drops of water before you close it up. Every few days, open each filter and check the seeds. When you are sure that no more will germinate count the seeds. If eight seeds germinate out of ten, the germination rate is eighty percent. If you put twenty seeds there and eighteen germinate, your rate is ninety percent. The seeds from some crops will germinate faster than others. Beans and squash are pretty fast, but peppers can take up to three weeks to germinate.

Once you have the germination rate, put it on your seed inventory and on the seed package. If the germination is low, it doesn't mean that you need to discard the seeds. Make a note and be sure to plan accordingly when you plant. Likewise, you might have better germination than you thought. That will save you seed when planting and save space in the flat, coldframe, or garden bed. Doing a germination test is also a good way to tell which varieties are more vigorous than others. You might find that the seed you save yourself or from a certain company germinates better than others.

Years ago, I had an interesting experience with bean seeds in the garden. It was right about April 25 and a really nice day. Obviously the weather was turning warm, but then, you never know at that time of year. I wanted to get a bed of snap beans planted and I was in a hurry. I had to go to Richmond to a meeting that day and didn't have much time to be in the garden. I had Provider snap beans from two different companies, intending to use the beans from the company that I thought had the hardiest seeds for that first planting. Well, I couldn't find those, so I planted the others. The soil was probably a little on the cool side. The beans came up, but there were bare spots and the yield from that bed wasn't what I'd hoped for. Once the weather warmed, I planted another bed of snap beans. This time I planted half the bed with seeds from the first package I'd used for the earlier planting and the other half the bed with seeds from the company that I thought would be the hardiest. When I had both packages together, I saw that the seeds from the package I had planted earlier varied in color and size. The other package had seeds all the same size and a uniform dark color. I had good germination from both packages in that bed when conditions were excellent. I imagine that the seeds in the second package would have done better in the cooler soil conditions of April. From that experience I learned to pay more attention to seeds. I also learned to not be in so much of a hurry.

Save Your Own

Saving your own seeds from one year to the next is the best thing you can do to insure the sustainability of your food supply. *Whoever owns the seeds controls the food supply*. If you save the seeds from the plants that do well under your growing conditions, you will have developed strains of those varieties unique to your garden. I've already mentioned growing open pollinated varieties. In order to make sure the varieties don't cross with one another if you grow more than one of each crop, learn more about isolation distances. A good seed-saving reference for your library is *Seed to Seed* by Suzanne Ashworth.

You might decide that seed saving is what you really like to do. In that case you could consider growing seeds on a large scale for sale through a catalog. The last chapter in *Sustainable Market Farming* is about just such a venture. Written by Ira Wallace of Southern Exposure Seed Exchange, that chapter explains some ins and outs of contract seed growing. If you are growing seeds to sell, you will also be interested in reading *The Organic Seed Grower* by John Navazio.

A website that you will want to visit is savingourseeds.org.[1] It is managed by Jeff McCormack who has a long history with seeds, including founding Southern Exposure Seed Exchange. Some of the information in this site is specific to the southeast, but there is plenty of good material there for seed saving in general. You don't need expensive equipment to save your own seeds. If you are growing beans, peas, and corn out for seeds to eat, you already have the skills for that. Tomatoes, peppers, and winter squash are easy because you are eating the mature fruit. The seeds can just be separated and dried. I leave them out on a plate until dry before storing them. Tomato seeds do need to go through fermentation by putting the seeds and pulp in a jar with water until the seeds separate out, but that is easy to do in your kitchen. Make sure that the plants you are saving seeds from are mature or the seeds won't be viable.

Varieties can sometimes disappear from the seed catalogs, just when you were getting to know them. That happened to me one year when I was growing lettuce to sell to area restaurants. The variety of romaine lettuce that I was growing had an open head, making it nice to rinse clean for my market. One year the seed catalog I had gotten it from didn't offer it. I still had seed left from the previous year, so I thought I'd try my hand at growing it out. I learned that lettuce plants can grow several feet high and are capable of producing a lot of seed. From just a small patch I harvested four ounces of romaine lettuce seed that summer.

Besides making sure you have the varieties you want that will grow in your conditions, saving seeds saves you money and contributes to the overall ecosystem in your garden. There have been major changes in the climate everywhere. Here in Virginia, winter and spring were unusually warm in 2012. As I write this, spring in 2013 is quite the opposite—cold and wet. Nature knows what is going to happen before we humans do. (We should pay more attention.) I wonder if the seeds know from one season to the next what to prepare for, just as animals know if they need to grow an extra thick winter coat or not. If so, the seeds that you save from your garden will be much better prepared for what's to come than seeds grown hundreds of miles away, or more. Whether you count your growing area in square feet or in acres, seed saving should be high on your list of priorities when you grow a sustainable diet.

10

Including Animals

A SUSTAINABLE DIET includes food that is grown on your farm or in your region and some of that might be animal products. Some land is better suited to pasture than to crop production. Including animals in your plan will expand the ecological footprint of your diet, since the land that grows the food that the animal eats needs to be considered. You can plan a diet of only plants, but you would be hard pressed to fill all your nutritional needs without taking supplements, which are not part of a sustainable diet. If that plan, which would involve a smaller area to grow your food, doesn't supply your needs, it is not a complete plan and needs to be expanded anyway.

In Chapter 3 I mentioned that if your diet only consisted of plants that you grew, in as small an area as possible, you would have to pay careful attention to getting enough calories, protein, and calcium. If you only ate plant foods, and even if you got enough calories, protein, and calcium from them, not only could some nutrients be out of balance, but the one nutrient that would still be missing would be vitamin B_{12}. Our bodies can store B_{12}, so if you had plenty of it in your diet for years, you would have extra that would carry over for quite some time, even years, if you stopped ingesting it. Eventually, though, you would run out. We happen to be at the top of the food chain and our bodies are adapted to a vast array of food choices. If we add just a little bit of foods from animal sources to our diet, we can fix these deficiencies.

In researching this nutritional information I discovered that the Recommended Dietary Allowances (RDAs) that I have long been familiar with have undergone some changes. The RDA is now part of the Dietary Reference Intakes (DRIs), which includes a set of four reference values: (1) RDA is the average daily dietary intake of a nutrient that is sufficient to meet the requirement of nearly all (97–98%) healthy persons; (2) Estimated Average Requirement (EAR) is the amount of a nutrient that is estimated to meet the requirement of half of all healthy individuals in the population; (3) Adequate Intake (AI) is based on observed intakes of the nutrient by a group of healthy persons and is only established when an RDA cannot be determined; and (4) Tolerable Upper Intake Level (UL) is the highest daily intake of a nutrient that is likely to pose no risks of toxicity for almost all individuals. The EAR for vitamin B_{12} for males and females age 14 and older is 2 mcg and that's what I'll be using for my calculations. The requirement is higher for pregnant and lactating women. The updated (1997) RDA for B_{12} is 2.4 mcg for age 14 and older.

Eggs were the first food I looked at for their B_{12} content. I found that the amount of B_{12} in eggs has dropped considerably from USDA's 1999 Standard Reference Release 13 that I found in *Nutrition Almanac*, to USDA's 2012 Release 25 that is available online.[1] I also consulted *Bowes & Church's Food Values of Portions Commonly Used* which has the 2008 Standard Reference Release 21. Food from animals that are on pasture has more nutrients than food from animals raised in conventional confinement systems. USDA's information is taken from conventionally raised animals, so I decided to go with the earlier B_{12} value which is 1.75 times the current (2013) available value. If you are interested in a sustainable diet, you would be eating eggs from chickens that have access to pasture. *Mother Earth News* magazine has taken an interest in the nutritional value of eggs and has found that eggs from hens raised on pasture may contain significantly more vitamin A, omega-3 fatty acids, vitamin E, and beta carotene,[2] but they didn't test for B_{12}. There is a 1974 British study[3] that shows the B_{12} content of free-range eggs to be 1.7 times that of eggs from factory farms.

I could meet my need for vitamin B_{12} with 2.6 large chicken eggs a day. Those eggs would also supply about 10 percent of my calorie requirement, 35 percent of my protein, and 7 percent of my calcium. If, instead, I consumed only 1 egg plus 1½ cups of whole cow's milk a day, I would meet my need for B_{12} and provide 15 percent of the calories, 40 percent of the protein and 45 percent of the calcium I need. Adding the milk really increases the calcium. Cow's milk has more than five times the B_{12} that goat's milk has. That's a puzzle to me, since there is not such a difference with the other nutrients. To get the same amount of vitamin B_{12}, I would need 2 eggs and 2.8 cups of goat's milk. If you are not a milk drinker, you could use milk instead of water to make hot cereal with your cornmeal or make yogurt and cheese with it. You might prefer to eat duck eggs, rather than chicken eggs. One duck egg provides 190 percent of the B_{12} requirement for an adult. Duck eggs are larger than chicken eggs, but comparing the same weight of duck to chicken eggs, duck eggs contain 3.7 times the vitamin B_{12} of chicken eggs.

Rather than eggs and milk, you could get your B_{12} from fish. Three ounces of light tuna canned in water would be enough. However, in a sustainable diet, locally harvested fish would be on the menu, not canned tuna. Three ounces of catfish would meet the B_{12} requirement and the same serving of trout has almost double that. I don't know much about fishing, so I won't be addressing that in this book. If fishing is a part of your life, or could be, include that in your diet plan. If you are getting your food from the water, you would want to work toward making sure the water is free of contaminants.

In a sustainable diet, when you are eating eggs and drinking milk, or eating cheese and yogurt, you also have to consider everything involved in bringing you those foods. These things are produced for you by the females. In order to have females, just as many males will be hatched or born. Those young males will become part of your diet, as will the females once they are past production. Otherwise, what would happen to them? They could just live out their days somewhere, I suppose, and that would leave an even bigger ecological footprint. I don't like to use the word slaughter, because that sounds like a vicious act. I don't like the word process either, because that sounds too industrial; but at some point, those animals will be killed to provide food for your table. When that happens, think of having shared your life with them, and now their energy will become yours. That energy is not going away. With that in mind, say a prayer of thanks before doing the deed and continue with respect for what you will receive. If you are buying your meat, rather than growing it yourself, buy it from someone who has the same values. When you grow and eat a sustainable diet, you become part of the food system. At the top of the food chain, there is nothing that is going to eat you, but maybe you could arrange to be buried in the woods at the end of your days to complete the cycle.

Chickens

Having a few hens in the backyard is usually the next step in homesteading after putting in a garden. A shelter can be cobbled together from almost anything, although some of the chicken house designs I've seen are quite elaborate (and expensive). Times have changed since I got my first hens and keeping chickens has become quite the thing. Local ordinances might have rules about what kind of chicken shelter you have, how many hens you can keep, etc. If your locality doesn't allow chickens, there is probably a group of eager citizens rallying to change that.

You can expect a hen to lay about 200 eggs a year—more in the first half of the year and less in the second half. I never put a light in my chicken house to push the hens to lay more, preferring nature to take its course. It seems that as the days begin to get the least bit longer, the number of eggs increase. Two hens would provide you with enough eggs to average an egg a day with a few extra for the year—it just wouldn't be an egg a day each day. Sometimes during the year you would be eating lots, and sometimes little or none. If you are concerned about B_{12}, that's okay. Since your body can store this vitamin for a long time, you don't need to ingest it every day. Those extra eggs you eat in the spring will provide the B_{12} you need later. Even if the hens are taking a vacation from giving you eggs, they will always need to eat. You can let them out in your yard to harvest what they can from the grass and insects, adding nutrition to their diet and yours.

Letting the hens out on grass is necessary with a sustainable diet, but controlling them is not as easy as one might assume. They can fly over a four-foot fence. If the area is large enough, however, they won't. At one time, I had all my hens in portable pens, inspired by reading *Chicken Tractor* by Andy Lee and Patricia Foreman. I moved the pens around the pasture to a fresh spot each day. If you do that in your yard, be prepared to have some low spots where the chickens have dug holes—even after just one day. Also, be prepared for chicken poop to be in your yard (and on your shoes). About Thanksgiving, my hens would all be moved to the chicken house with a yard enclosed by a six foot high fence. They'd go back out to the chicken tractors in March when things started to green up. We have fewer hens now, and I like them to not be so confined, so I don't use chicken tractors anymore, but they are a great way to manage chickens in limited space.

Now, our hens are sheltered by the chicken house all year, but they have more room to roam, even in the winter. In the process of doing other fencing projects, it turned out the fences we put up allowed the chickens to be free to wander everywhere but the yard and garden. That's something to keep in mind when making your permaculture plan. I'll talk more about fencing in Chapter 12.

I buy old farm books when I find them. They provide great information on how things were when every farm was small and diversified. I have to sort out the good information from the not-so-good—such as using creosote on the chicken roosts to prevent lice. (In case you don't know, creosote is a banned product now.) One book[4] told of the necessity to raise growing pullets (young hens) on pasture, in addition to feeding grain. It was suggested that one acre could pasture 600 pullets (about 74 ft² per pullet), reducing the cost of raising them by five to ten percent. According to Joel Salatin in *Pastured Poultry Profits*,[5] it is possible to reduce grain consumption by 30 percent when the chickens are on pasture, with no drop in egg production. When you buy chicks or hens, it makes a difference where they've been before you get them. You want ones that have been raised on grass, or have had parents raised on grass. With no other choices you could order from a hatchery, and I suppose everyone should do that once in their lives. It's an adventure to get a box of chicks in the mail. Locally, you could find poultry advertised on Craigslist, the local newspaper or swap news, or inquire at your feed store. They always know who's doing what.

Getting the chickens out in the grass allows them to eat the grass, weed seeds, and bugs; which is healthier for the chickens, the grass, and for you when you eat the eggs. I find my chickens love to dig around in piles of leaves and along the fences. They are known to keep down the tick population. For good egg production, foraging like that isn't enough. If I grew enough grains here, I could feed them that, but that's not the case. I buy organic corn, wheat, and oats and grind my own feed. The mix I feed my chickens is 60 percent corn, 20 percent wheat, and 20 percent oats. I have been doing this since 2000 when I became concerned about genetically modified ingredients, especially soy, in the chicken feed. I keep ground oyster shells available to them for extra calcium. (You can buy ground oyster shells for this purpose; I don't grind them up myself.) Although, since they are out scratching in the ground, most likely they are picking up calcium there. You can also supply calcium in their diet by feeding eggshells—crushing them first so that the chickens don't get the idea that they can just eat eggs. I feel that by having my hens free-range during the day, they are balancing out their diet on their own, picking up needed grit, vitamin D (from the sun), and other things that hens kept in confinement need to have provided in their feed. I could put out less feed and keep more hens to get the same number of eggs, which is easier to think about now that I'm not selling eggs. Sometimes farmers who grow grain will sell it to you out of their bins, right at the farm. Make sure to ask about their growing methods and seed source.

My grain mix does not have enough protein for raising chicks. The chicks grow and feather out much faster when I give them worms, in addition to this mix. I kept a worm bin going in my house for many years to drag around to the classes I taught. It was handy for putting food scraps in during the winter. By spring, the bin would be at peak production, just in time to harvest some worms for the newly hatched chicks. I use an incubator for hatching eggs, but having them hatched and raised by a broody hen would be better. Studying the chicken feed issue has always been on my to-do list, but a little further down than the other things I've talked about in this book; so I haven't considered it as much as I would have liked. Fortunately, my friend Harvey Ussery has. He has tested all sorts of things, including breeding black soldier fly larvae to feed his chickens, and he tells you about it in his book *The Small-Scale Poultry Flock*. The old farm books I've found often have lists and descriptions of all sorts of things to feed chickens and other animals.

When I had fifty layers and sold eggs, I would hatch out at least fifty chicks each year. About half of those chicks would be roosters, which would be taken for meat by 12 weeks of age. The laying flock would consist of 25 hens in their first full year of laying and the same number in their second year. By fall, the new pullets would be starting to lay when the oldest hens were slacking off. That's when we took the old hens for meat. I kept close financial records and broke even money-wise with the egg sales. Our profit was food for our table in the form of meat and eggs.

Although hens will lay eggs quite well without a rooster being around, a rooster is needed to fertilize the eggs in order to have chicks. One rooster for every ten to fifteen hens is a guideline. They can be noisy

with their crowing. Lucky for us, our neighbors enjoy the sound. It is part of country living. Roosters are usually banned from city and sub-urban lots—all the more reason to work on building community so that someone is keeping roosters with their hens and raising replacements for you if you need them. Even if you have your own rooster, predators abound and can wipe out your flock, or part of it, in short order.

If you want to have ducks you can raise them without a pond, but you should still provide some water for them to splash in. I don't have experience with ducks, but Harvey mentions them in his book, as well as guineas, and turkeys. A three ounce serving of roasted chicken or duck provides about 15 percent of the daily vitamin B_{12} requirement.

Dairy—Goats and Cows

When we first moved here in 1984 I bought two Nubian goats for milk. We didn't have any fenced pasture yet, so I tied them out each day, keep-ing a water bucket handy. We fenced a very small area so that if I was gone for any length of time during the day they could stay there. We had a shed that I put them in each night. I was worried that dogs might come along and bother them when they were tied out. After a year and a half we fenced some pasture. That was a lot nicer than putting them on the end of a chain each day. In some countries it is common to tether goats and cows and that's how the grass is controlled along the roadsides. I could only imagine what people would have had to say if I had tied my goats along the road. Come to think of it, it would be easier than mow-ing the ditch.

I made a milking stand from scrap wood we already had. For what-ever else I needed for milking, I used what I had around the house. Goats are easy enough to put in the back of a pickup truck, providing you have sides on it. They can even jump in and out without a ramp. Some people move them in a van or car. You would need to consider moving them to take them to a buck to get bred, unless you keep a buck yourself. Be aware, a buck can be smelly. If you bring goats home as dairy animals, make sure you know how you will get them bred when the time comes.

Although books I've read show that one goat could give you up to four quarts of milk a day (one gallon), I believe I got a quart of milk a day from one goat, with two milkings a day. The second goat was not always producing when the first one was. Being tied out each day, rather than free to wander the pasture, could have affected the milk production. The condition of a pasture will always affect how much milk a dairy animal will give. We hadn't done anything to the pasture—there was already grass there and we let the goats eat it. In order to have milk, a goat would have to have given birth. A six-month old goat kid, taken for food for your table, could yield 30 pounds of meat and bones.[6] We only had dairy goats for three years, in the midst of fixing up an old farmhouse, raising children, etc. Goats like to browse on bushes and trees more than grass, and are often used to clear areas of brush and poison ivy. If you move to someplace that has an overgrown area you'd like cleared, goats just might be the answer. A temporary fence of livestock panels[7] could control their movement and is fairly easy to move.

For seven years we kept a milk cow. She was too big for the goat shed, so we fixed a place in the barn for her. The goats hadn't been in the barn because we were using it for other purposes at that time. I penned the calf separately at night and milked only once a day in the morning. The calf took the rest of the milk during the day. If we had to be gone overnight, the calf took all the milk. The calf was taken for food for us at about ten months. In our area we have a local butcher who comes to the farm, kills the animal there, and takes it back to his place to age in his cooler for a few days, until he cuts it into pieces for the freezer. Legally, that's okay, since we are consuming the meat. If we were to sell the meat by the cut, it would need to be federally inspected. Someone, or a group of people, could buy an animal from the farmer and have the butcher process it for them, and that would be legal because it would already be their animal. In the not-so-distant future, I believe traveling on-farm processing units will become the norm. If you were receiving 0.7 gallons of milk a week (1.5 cups/day) from a cow producing a gallon a day, you would be getting ten percent of the production. Using conservative estimates, if the calf was taken for meat at ten months, you would get about twenty-three pounds of meat for the year, plus bones for broth. That would be about seven ounces per week for your diet.

Having a cow and milking once a day worked well for us. The biggest problem was getting her bred. There is a small window of time when a cow is in heat. If you miss that, you have to wait until the next time. She will give you clues, such as mooing constantly. You could go out and lift her tail to find physical clues, but with only one cow, if you are new at this it is hard to tell. If you had two cows, they would be jumping on each other. I only knew one other person who kept a family cow and he sold it about the time I got mine. There was a man whose business it was to go to farms to artificially inseminate (AI) cows, but he retired about that same time. In the end, Tommy, the dairy farmer down the road, offered to come when I needed him to do the AI. I would call and he'd promise to get over that morning, but sometimes things happened and it would be after dark when he came. It didn't always take and he would need to come back on the next cycle. Tommy never seemed to mind, something for which I am forever grateful. We always had a nice chat when he came, catching up and talking about things on our farms—our very tiny one and his very large one.

If I had known someone with a bull, I could have hauled the cow to the bull, or vice versa, if I had wanted to have a bull in the pasture. Actually, there was someone who offered to bring his bull over for a month to make sure our cow was bred. However, this bull had quite a spread of horns and, not being familiar with bulls, I wasn't sure I wanted the responsibility of having that animal at our place. As far as hauling the cow, she was too big for the back of the pickup, so I would have had to find someone with a trailer to do that. When keeping large farm animals, it is also important to find a veterinarian who will come to the farm. It is easier to keep a milk animal if there are others doing the same. A community will evolve to provide for your needs.

Water is a consideration. Cows drink a lot of water and, if there was a winter storm threatening, I would fill six five-gallon buckets of water in case the power went out. Our water comes from a well and has an electric pump. With the cow and calf, those six buckets would last only one day. There is no faucet in the barn, so all the water was run out there with a hose. From December through February, the water was carried out in buckets, since the hose would be frozen. If I was to have a cow again I would seriously consider having running water in the barn and make sure we had adequate water storage in case of power outages, which seem to happen more frequently now, and not only with winter storms.

Besides milk and meat, cows can provide draft power. Rather than being taken for meat at a young age, the bull calves become oxen. An ox is a castrated male trained to do work and can be from any bovine breed. Beyond their working days they will become food for the table. It takes a lot of training when they are young, but once trained they can be easily managed by the right person—the drover. Our place is small enough that we don't need draft power and we don't have enough pasture for draft animals, but we have raised a drover. Luke became acquainted with oxen when he was five and, now that he's grown, has his own teams. The skill of training these animals for work is being kept alive by people like him and those in places that depend on draft power around the world. Interesting things have been happening in Cuba. Cuba was dependent on the European Socialist Bloc and the Soviet Union for many things, including tractors and agricultural chemicals. When that support system fell apart in 1989, Cuba had to find other means to keep its agricultural systems going. Those systems changed to include organic urban agriculture initiatives and increased use of draft power on farms.[8] Steers had to be taken out of the food system to train as draft animals, more drovers had to be trained, and pasture land needed to be increased on farms that had previously been managed with tractors.

Ideally, the milk and meat in a sustainable diet would be produced on pasture only. It is healthier for the animals and for the people drinking the milk. More and more, dairies are going to organic and grass-fed systems. Traditionally, whey (from cheese making) and skim milk (from butter making) was fed to chickens as the animal protein that they needed. Dairy animals and chickens go well together on a farm. A three ounce serving of ground grass-fed beef is about 80 percent of a day's requirement of vitamin B_{12}. The same size serving of goat meat is 50 percent of the requirement.

Swine

If you have dairy animals and make cheese and butter, you might want to consider a pig. Whey and skim milk are excellent foods for pigs. Homesteads have long depended on pigs for a variety of things. Besides meat, there was fat to be rendered into lard for cooking and for soap making. Traditionally, the meat was preserved by salting and smoking. We had pigs at our place one year when one of our children raised them for a 4-H project. We found that they ate a lot of corn. At first we bought pig feed, and then switched to whole corn. We were new at this and weren't set up to provide any homegrown food on a regular basis. We could have just let the pigs run in the woods and eat acorns, but back in the day when people did that they would still pen their hogs and fatten them on corn for a few weeks before they butchered them.

We started with two feeder pigs (weaned piglets) and we put them in the pen we had earlier used to keep the goats in, since the goats had a fenced pasture by that time. That pen had a tall metal gate, the kind you might find on a chain link fence. It was something we had acquired for free. The gate hinges were of the lift off variety, with barrel sleeves that drop onto pins on the gate post. The gate worked quite well for the goats. One day, our children threw some food scraps from dinner into the pen, and they landed right in front of the gate. In their enthusiasm to get the scraps, the pigs rooted a little too hard and lifted the gate right up and off those pins, setting themselves free. We all got our exercise that evening getting them back in. We put a block of wood in place above one of the pins, so that couldn't happen again. We also made a food trough that we could easily put food scraps in from outside the fence. It is good to start out small with homestead projects until you really know what you are getting into. That was the only year we raised pigs. In recent years we did some more fencing, and I had that episode in mind when we decided to fence the barnyard.

Luke was born the summer we had the pigs. I was also milking the goats then. For the first week after he was born, and occasionally after that, the pigs received the goat milk. The last month or so, they had outgrown their pen and we let them into the pasture. When the come-to-the-farm butcher was cutting up the meat from the first pig to package it, he was so impressed at the quality of the meat that he bought the other pig himself. He caters events with barbeque and was happy to get such good pork.

Besides dairy and pasture, root crops, especially Jerusalem artichokes, are great feed for pigs. Let into the root patch, they can do the harvesting themselves. Winter squash and other garden vegetables make good pig food. A person growing for the markets will find the pigs willing to chow down on all the leftover and cull produce available. A three ounce serving of pork contains about 40 percent of your B_{12} requirement for a day.

Rabbits

We had rabbits for about six years—another 4-H project. We found directions for making a wire rabbit cage in *Integral Urban House*. Out of print for many years, I am glad it is available again. I like that rabbit cage design because it has a hay manger between two cages. We followed the suggestion in the book to have the rabbit cages overhanging the chicken yard. The chickens scratched through their droppings, making finished compost on the spot. Alternatively, you could locate worm bins under the rabbits to catch the droppings, and feed the worms to the chickens. Having the rabbit/chicken set-up is a good one, but I would prefer to have the rabbits on the ground, which presents a quite a different set of management decisions. The Salatins at Polyface Farm[9] in Swoope, Virginia, have been doing that for many years. If you are breeding rabbits, you would need a buck and one or more does, with separate cages for each, plus one or more cages for the weaned young ones to grow to butchering age, which is two months. A doe will give birth a month after breeding. The average size litter is eight bunnies.

Rabbit food you buy consists of pellets made predominately from alfalfa. If you were to raise all the food for your rabbits, it would take 9 pounds of alfalfa hay and 60 pounds of fresh greens (garden debris, weeds, produce scraps) to produce one four-pound (live weight) fryer.[10] Your rabbits would have to be accustomed to eating that much, since about 15.4 pounds of pellets will get the same results. Even if you don't grow everything, you can supplement pellets with alfalfa and greens from your garden. Alfalfa is a suggested food for all the animals. It could take the place of red clover in your garden rotation (see Chapter 8—Garden of Ideas), possibly staying in for an extra year, before the next crop replaces it. The average yield of alfalfa hay in the US for 100 ft² is 14.9 pounds.[11] Feeding the alfalfa to your livestock means that you wouldn't have it to put in the compost pile, but now the animals are part of the cycle and their manure would go to the compost, which would go back to feed the alfalfa, as well as the rest of the garden. A three ounce serving of rabbit meat contains three times the daily requirement of vitamin B_{12}.

One advantage of raising rabbits and chickens is that taking the meat for the table can be done one animal at a time, as needed, without canning or freezing. Even if you butchered a litter of rabbits at a time, you could fit them all in the freezer space of your refrigerator—a big consideration when living with minimal fossil fuel. Manage your rabbit herd so the next litter is not ready for the freezer until there is room for it. A sustainable diet is plant-based, but not plant-only. Using animal products to round out your nutritional requirements also means using the animals to harvest plants in as sustainable a manner as possible. Land can be kept in permanent pasture as long as it is not overgrazed by stocking too many animals there.

I mentioned that goats can clear land of brush for you. If you don't need land cleared, recognize their brush-eating preference and harvest it for them. When we had bamboo that needed to be cleared, everything I cut from the bamboo poles went to the goats. They considered it a treat. I tried to do that in winter or early spring before the pasture was growing again. The bamboo poles were kept for garden projects. Pigs also can be used to clear land. Let out on enough pasture, they won't do too much damage; but penned in a smaller space, with maybe some stumps, they will root them up. Properly fenced, they could rotate through garden space after Jerusalem artichokes, carrots, beets, or mangels grown for them, leaving the area cleared and fertilized. *Properly fenced* is the key for that.

There are certainly more nutritional requirements to consider than just calories, protein, calcium, and vitamin B_{12}. *Nourishing Traditions* by Sally Fallon is a good reference for learning more. I've shown you some ways to include animals in your food production circle. Often things can be added to a system that will enhance the whole system, and rather than overloading it will make the best use of all the available resources. Enhancement is the goal.

11

Food Storage and Preservation

HUMANS HAVE BEEN STORING FOOD to eat later since long before electricity became a part of our lives. Unfortunately, today many don't know how to manage that without electricity. A sustainable diet uses the least fossil fuel to get food from the garden to the table. Folks are trying to be more self-sufficient, as evidenced by the explosion of new gardens; which makes food preservation, especially canning, a popular topic. It reminds me of the back-to-the-land movement in the 1970s. The Bicentennial in 1976 spurred even more interest in the home arts. Those were the years I was starting to garden and learning to can. Both Kerr and Ball, along with a few other companies, were producing canning jars and I still use some commemorative Bicentennial canning jars I bought at that time. (Bicentennial canning jars are kind of silly when you think about it, since canning was not a preservation method in 1776.) The magazines on the newsstands had articles about canning and I could buy everything I needed to get started, so that's what I did. I had both the Ball and Kerr books for home canning and freezing. There were also books that told about other methods, such as root cellars, making cheese, and fermenting. *Stocking Up* came out in 1973 and was on its tenth printing by the time I bought my copy in 1975 and *Root Cellaring* was published in 1979. My biggest inspiration to go beyond

129

canning was reading *Home Food Systems*, which came out in 1981. After checking it out of the library many times, I decided I needed to buy a copy. The sticking point for me to go beyond canning was that I didn't know anyone who was doing it. It is always better to have a flesh-and-blood mentor. Lacking that, even attending a talk or workshop would have helped. Today there is once again a self-sufficiency movement, but it is not necessarily back-to-the-land like before, when living in the country on a farm was the goal. Perhaps now it should be called bloom-where-you-are-planted, what with the current interest in urban farming.

Times certainly have changed. *Mother Earth News*, a homesteading magazine that had its beginnings in 1970, has an online presence with bloggers, such as me. They sponsor the Mother Earth News Fairs in Washington, Pennsylvania, and Kansas, where you can go to hear the people you have been reading about—once again, such as me. Communities are sponsoring workshops and programs that help people lead more sustainable lives. When you first start out learning new things, it can seem hard and, at first, not very productive. I'm here to tell you that it gets easier. In fact, with a little knowledge, it doesn't have to be that hard to begin with. Any change brings chaos to some degree, but once it becomes your way of life, things stabilize. I will share what I've found to be helpful.

Make Use of the Space You Already Have

With staple crops grown for a sustainable diet—potatoes, sweet potatoes, garlic, onions, winter squash, grains, peanuts, and tree nuts—you don't have to dig a root cellar or buy special equipment to store them. Growing up in northeastern Ohio I thought every house had a basement, but I learned that wasn't so when we moved to Virginia. Since the frost line is not as deep here, the house foundations aren't as deep; so mostly there are just crawl spaces, unless your house is on a concrete slab, and then there is no chance for storage at all. I didn't know anyone storing food in a cellar in Ohio, other than canning jars, but the potential was there if you had one. Before you think about under-the-house storage, however, take a look at what is available in the house.

A spare room or closet can become a food storage area. Turn off the heat to that room and maybe open a window a little. It doesn't have to be too cold, just cold enough (about 50°F). A closet on an outside wall might be cool enough already. We never had any spare rooms or extra closets, so I never made use of that idea. Space was pretty tight with a growing family. One winter, while retrieving a bread pan, I discovered that the cabinet beneath the kitchen counter in the northeast corner of the room was exceptionally cool. A check with a thermometer showed the temperature to be about 50°F. My first thought was that I had better keep that cupboard door closed. My second thought was—why am I storing baking pans here instead of potatoes, winter squash, onions, and garlic? It was easy enough to clean out that cabinet and add produce stored in baskets, cardboard boxes, and plastic tubs. It would also be a good place for grains, beans, and nuts.

We had been in the house for enough years that it was time to re-think what was stored in all my kitchen cabinets. Even if you don't have children being born or leaving home, your lifestyle gradually changes and it is good to do a major clean-out now and again. There were things stuck back on the shelves in the bottom cabinets that I never used—because they were stuck in the back of the bottom shelves. When our time opened up for a kitchen project, my husband took out those shelves and replaced them with pull-out shelves he built. Forget major appliances and other improvements, for me that was about the best thing he could have done. During that time we took the opportunity to paint, inside and out, the plywood cabinets that were standard in about 1960 when they were put in.

Having pull-out shelves means that everything is readily accessible. No more unused things hiding in the back. The space that was to become my potato/squash/onion/garlic storage area was behind two cabinet doors to the left of the sink. We made a plywood wall to block that space off from the under-sink area. The option exists, if we get ambitious, to put a vent in the bottom and one near the top to let air flow through from the outside. In that case, I might insulate the doors. For now, ventilation occurs when the doors are opened and closed as we get things from the cabinet. In the winter, the temperature in that cabinet is colder than the room, usually around 60°F or a little lower. I'm not sure it ever consistently stayed at 50°F—maybe it was 50°F that day to get my attention. Venting to the outside would increase air flow and lower the temperature. While I was in my take-a-new-look-at-everything mood, I cleaned out the packed junk drawer just above this new produce cabinet and now we keep bread there. Hoosier cabinets from days long past had tin-lined bread drawers and that's where I got the idea. The bread box moved from the kitchen counter to the pantry where it now holds extra light bulbs. As for all the junk I cleaned out of that drawer—it was sorted and put to good use elsewhere. No one has missed it.

The new pull-out shelves all have solid bottoms, except for in the produce area. For that we used a heavy metal grid, capable of holding the weight and allowing air to flow through. In the produce cabinet, I can fit a bushel of Irish potatoes and a bushel of sweet potatoes, plus winter squash, onions, and garlic. If I have more than that, it can go in plastic boxes in the crawl space or a cool closet elsewhere. Sweet potatoes and winter squash can even be stored under a bed, although the space under our beds has always been full of other things.

We do not have an electric dishwasher, preferring to wash our dishes by hand. If we had a dishwasher, we might not have room for storing produce in an under-the-counter cabinet. If you are looking for more storage space under the counter and think that the dishwasher is just in the way, but that it's too much trouble to take it out, use it as storage, freeing up other cabinet space.

Crawl Space Root Cellar

Sweet potatoes and winter squash are happy being stored at 50–60°F at 60–70% humidity. Irish potatoes are happier being stored a little cooler and at 80–90% humidity. If they are too warm and dry, they will sprout earlier. I fill the box in the cabinet in the kitchen with potatoes we are going to eat and keep the extras in the crawl space under the house in ten gallon plastic boxes with air holes drilled in the boxes and the lids. They don't go in there until October. Until then the potatoes are stored in wooden bushel baskets with newspaper on top to keep out the light, or stored in paper grocery bags and kept either in the house or in a shed. The baskets and bags allow the potatoes to breathe. Before storing them in the crawl space, I sort the potatoes carefully, discarding any that have gone bad, and counting out how many I need to plant for the next spring, keeping them separate. The furnace is in the cellar—keeping the cellar a little warmer and often drying it out—both bad for the potatoes. Conditions in the crawl space are better suited. Even when the temperature dipped into the teens, the temperature in the crawl space never dropped below 48°F. You can monitor the temperature with one of those remote sensors that operate with or without wires. If it has a wire, you can take off the baseboard trim in the house and drill a hole in the floor next to the wall, put the wire through, then replace the trim, making a groove for the wire, allowing you to check the temperature of the crawlspace from inside the house. You might want to do this in a utility area or a closet. I know someone who ran a wire that way to put a light in his crawl space. The cord came up through the floor into a bedroom right beneath an outlet, where it plugged in. If you do drill through the floor, make sure you know where it will come out below. You don't want to drill through an electric line or water pipe.

The extra onions and garlic are either braided or in mesh bags and hang on nails I put in the joists in the crawl space. Like the potatoes, they aren't put there until October. Until then they hang in the barn where I put them after the June harvest. Sometime in July and August, I sort the garlic and set aside what is to be planted in the garden in late October. The multiplier onions are managed the same way and planted out with the garlic. The extra sweet potatoes and winter squash are usually stored in the house, if I have room. The sweet potatoes to plant the next year go under the house, clearly marked, so I don't use them by mistake. Once fall comes, these things could be stored in your attic, if it is cool and accessible. If it is cool in the winter, it is probably too hot in the summer. With the super-insulated houses of today, I'm not sure how attics accommodate produce. To get into our attic we have to get out a ladder, so we don't store produce up there.

Obviously, these storage temperatures cannot be maintained throughout the summer using natural methods. Our house is not air conditioned. We rely on shade trees, cross ventilation, and ceiling fans for cooling. In spite of that, the potatoes, sweet potatoes, winter squash, garlic, and onions survive very well with the plan I've told you about.

Pantry

We have a room off the kitchen that we used to call the utility room, but now it's the pantry. When the house was built—sometime between 1900 and 1910—the kitchen was smaller and this room was a room off the porch used for storage. The family who lived here kept two cows and had chickens, selling butter and eggs each week. I imagine some of that was kept in this room, although the butter might have been kept in the cellar that was dug out by hand under the kitchen in the 1930s. About 1960, plumbing was added to the house and the kitchen was enlarged to include the porch, making that storage room part of the new kitchen. By 1984, when we bought the house, washer and dryer hookups were there. Our washing machine is still there, but we have given away the clothes

dryer since we rarely used it. We dry our clothes on the line outside or on a wooden rack inside. By removing the clothes dryer, we made space for fermentation crocks and for baskets and crocks of peanuts, hazelnuts, and black walnuts.

This room needs strong shelving. Besides the crocks and baskets on the floor, it holds large and small jars of grain, dry beans, honey, and dried food. It is cool, dry storage. If there was room, I could put the canning jars here, but I keep those in the cellar. Although the furnace sometimes dries out the cellar, other times it's damp there—not my preferred place for beans, grain, nuts, and dried food, even if they are sealed. Half gallon canning jars are good for storing grains and beans in. The rubber gasket on the canning lids keeps out air and moisture. If you see quart and pint canning jars for sale in a store, but not half-gallons, inquire about having the larger jars ordered for you.

Many things that used to come in large glass jars are now packaged in plastic. My aunt owned an ice cream store and would bring me boxes of gallon and half-gallon jars she had emptied of toppings for the ice cream. At that time, it seemed there was a never-ending supply of those jars. Today those toppings come in plastic containers. You could use those since, obviously, they are food-safe.

Food-safe plastic buckets are good storage containers, especially for grains and beans in large quantities. Bakeries and restaurants get ingredients in them and often make them available to whoever wants them for a small fee or for free. All you have to do is ask. You can find one-gallon and five-gallon buckets this way. Food-safe plastic buckets are very useful on a homestead, both in the pantry and in the garden. One winter when I had some extra time, I called around to see if anyone had any buckets I could have. A donut business said that I could come and get some one-gallon buckets for free. When I arrived they also offered me a whole pallet of five-gallon buckets full of fudge topping that had proven too thick for their machines. I don't know how long it had been stored there, but they seemed happy I was taking it off their hands and freeing up space. You are probably familiar with the phrase—be careful what you ask for because you might get it. I thought that I would just dump the topping in the compost (it was unsweetened; we tasted it) and clean out the buckets. I soon learned that it would be better to wait until summer and let that stuff soften in the heat a bit so it wouldn't be so hard to get out of the buckets. Even then it was a chore, and a messy one at that. It took me more than a year before I got around to emptying all those buckets, but I haven't had to go looking for buckets since.

Cooling Cabinet

There was evidence of an old furnace having been in our pantry, but it had been taken out when the new one was put in the cellar before we moved here. Left from that furnace was a hole in the floor and a larger hole about four feet up for a flue going into a chimney. I always thought I'd enclose an area in that room to make a cooling cabinet, such as the one I had read about in *Home Food Systems*.[1]

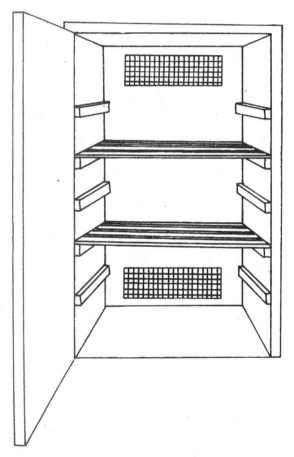

Eventually, as that room made the transition from utility room to pantry, by getting rid of the unnecessary things and storing more jars of homegrown grain, beans, honey, and dried food, I realized that I could just uncover those holes, screen them to keep out critters, and have the whole room vented. There wasn't enough room for a separate insulated closet or cabinet in there, anyway. Originally there was a window in the pantry, but that had been taken out when we put an addition on the house. From my research into pantries, windows were common in order to vent and cool the room. Now I had venting again.

The cooling cabinet I had read about was a description of something that had apparently been common in times before refrigeration in southern California, where the days were hot but the nights were cool. Cool air from under the house was drawn through the cabinet by natural convection to a chimney.

That's where I got my inspiration for the produce cabinet in the kitchen and why I had thoughts of putting in lower and upper vents.

Fermentation

When living with a diminished supply of fossil fuel, fermentation is a wonderful way of preserving food. Living with a diminished supply of fossil fuel, by the way, has nothing to do with living a diminished life. It is a life of new and exciting opportunities and Sandor Katz is the biggest inspiration I know to help you realize those opportunities. I had read about fermenting vegetables one jar at a time in *Nourishing Traditions*, but hadn't gotten around to doing it. When I read *Wild Fermentation* by Katz, I just had to try it. The first sauerkraut I made was in a quart jar using the directions in *Nourishing Traditions*. The directions called for shredded cabbage, salt, and whey. You can obtain whey by draining it off yogurt. The whey is added to jump-start the fermentation process, putting folks at ease who are not used to leaving food sit out unrefrigerated. I soon became comfortable with not adding whey. Natural lactic-acid producing bacteria are already present on the cabbage leaves. The salt provides the right conditions to get things going. Although cabbage is the base for sauerkraut, you can add other things, such as radishes, apples, carrots, and onions. Sandor Katz gave a demonstration at the community college in 2007, a public event sponsored by the college sustainable agriculture club, and had my students chopping an assortment of vegetables to accompany the cabbage in the ferment. Add salt, mix well, and pack it into jars is all it takes. I could have just talked to my students about fermentation, but having Sandor there really moved everyone forward along that path. Take a look at his website at wild fermentation.com and let Sandor inspire you.

I've made sauerkraut in quart jars, old-fashioned crocks, and Harsch crocks (a brand of crocks that have a water trough around the top). The quart jar method is easy and is great if you have a limited quantity of produce to work with. One head of cabbage makes a quart. I enjoy making larger quantities at a time. When I made it in the old-style wide crocks I realized why people were so quick to abandon the crocks when refrigeration became the norm. You have to know how to manage them and regularly pull the scum off the top, clean the lid (plate and weight) and replace it. The scum looks pretty bad, but it won't hurt you. The trough around the top of the Harsch crocks is filled with water to help keep out air. It is contact with air that leads to the scum. Check the trough each week and add more water as necessary.

Using plywood and wheels I made small rolling platforms for my crocks so that I can move them around easier. These platforms are like the ones you would put plants on, but sturdier. The fermented vegetables keep for quite a long time in the crocks. Some instructions say to take the fermented food out after six weeks or so, but it is okay to leave it in all winter if you want, as long as it tastes alright. I have eaten sauerkraut that has been in the crock for more than a year. The taste changes as it ages. You may prefer either a young sauerkraut or an aged one. Taste it regularly and enjoy all the flavors. If you need the crock for another food project, put what you have fermented in jars. If you want to stall the fermentation at that taste, store the jars in the refrigerator.

I look forward to making dill pickles—called sour pickles in *Wild Fermentation*—from my cucumbers, garlic, and dill. I have one-gallon and two-gallon jars that I use. If possible, I pick all the cucumbers at just the right size to ferment them whole. If they get too big, I quarter them lengthwise to put in the jar. I put 3–4 pounds of cucumbers in a gallon jar, add 6 tablespoons of sea salt, a few heads of dill, the peeled cloves of several bulbs of garlic, and then fill the jar almost to the top with water. Since I have grapes, I put some grape leaves in the bottom of the jar before I start. They are supposed to keep the pickles crisp. To keep the cucumbers from floating to the top, I add a smaller jar that just fits the opening of the large jar, or fill the space with a water filled zip-lock bag. It keeps the air out and keeps the cucumbers submerged. There is no cooking and no vinegar involved. I leave this on the kitchen counter—deli style. After a couple weeks we begin to help ourselves. That's when I start the second large jar. Once summer wanes and the pickles are about as tart as I want them, I put what's left in quart jars and store them in the fridge. I used to can pickles with vinegar and thought I needed to put up enough for all year. Now we celebrate the seasons. Once the pickles are about gone, the sauerkraut is coming along. There is a season to everything. I could put the sour pickles in a large Harsch crock and have more and for a longer time, but then I wouldn't have so much fun seeing those large glass jars of pickles in the kitchen.

I salt down snap beans in a crock to preserve them by mixing French-cut beans with sea salt and packing them in a crock. I have a bean frencher to slice the beans, but you could slice them lengthwise with a knife. Pack everything in tight and in a few days the water drawn from the beans will be present as liquid in the crock. I usually can the snap beans produced early in the season, and preserve the ones that come later in the summer in the crock. If you don't have enough to fill the crock when you start, you can add to it as the beans come in, which is nice if you have a small garden and only harvest a pound at a time. (You can also add cucumbers to the pickle jar/crock after it is started.) Before eating, the beans need to be soaked in plain water for an hour or two to remove excess salt. They will also need to be cooked before eating—steaming will do. You could take them from the crock, chop them smaller, and put them in soup to cook just as they are, without adding salt to the soup. Once when we lost electricity for several days due to a hurricane, I was able to preserve my bean harvest by salting.

We make mead (honey wine) from our honey and grapes, fermenting it in gallon jugs stored on shelves in the pantry. Once you learn to do these things, you will find more ways to use your new skills. Fermented food is good for you. People have told me that they have cured their acid reflux problems by eating naturally fermented sauerkraut. Can you imagine how much money the drug companies would lose if everyone did that? Naturally fermented food has a multitude of health benefits, including fighting cancer.[2] In preserving food with fermentation, you are participating in your own health care. I keep the crocks in the pantry or any odd corner that I can find. The warmer the spot, the faster the fermentation. If you want things to slow down, find a cooler spot.

Canning

Canning takes equipment, know-how, time, and fossil fuel. It also heats up your kitchen in the hottest time of the year. Nevertheless, that's the method of preserving that most people know about, besides freezing.

I won't be talking about freezing because it requires continuous electricity to keep the food preserved, and if the electricity goes off the food would be lost. Recognizing the need to educate the public about safe methods, the National Center for Home Food Preservation was established in 2000. If you want to learn more about canning you can find the USDA *Complete Guide to Home Canning*[3] as a free download on their website. This book is also available to order as a print copy[4] (not free). Available as a book to buy is *So Easy to Preserve*,[5] which has information on canning, freezing, and drying. Information from this book is available as fact sheets to download for free. Every homestead kitchen needs a good comprehensive reference and I highly recommend these books, particularly if you are just learning. I'll limit my remarks here to how to use less energy while canning.

You will discover that there are two types of canners. A water bath canner is a big pot with a rack in the bottom. You fill it with enough water so that the boiling water comes an inch above the jars. This type of canning is for high-acid food, such as pickles, fruit, jams and jellies, and tomatoes. A pressure canner is more expensive than a water bath canner, requires less water, and is for preserving low-acid food, such as vegetables and meat. When you open a jar of low-acid canned food (vegetables and meat) you should boil it uncovered for ten minutes before eating, to guard against botulism.

Pressure canners are often advertised by how many quarts they hold. That means how much they hold if the pan was filled with liquid. A six quart pan holds six quarts of liquid, not six quart jars to can. I mention that because I've seen store displays with canning equipment where the pressure canner on display was actually a pressure cooker and would only hold a few pint jars. That's okay if you want something to pressure cook meals with and occasionally put something up in pints, but if you are doing any amount of canning, buy one that will hold at least seven quart jars.

The easiest way to get started is with water bath canning. When I first started canning I put up pickles, jam, peaches, applesauce, and tomatoes using my water bath canner. After a few years I bought a pressure canner. I grew potatoes and canned some because I wanted to preserve them for the winter. I didn't know how to store them at the time. Potatoes and carrots take longer than other vegetables to can, using more energy. I don't do that anymore. My winter potatoes come from stored supplies and my winter carrots are harvested fresh from the garden.

Besides snap beans, I can tomato products, particularly tomato soup, to have as a convenience food—heat and eat for lunch. I make all our spaghetti sauce. I used to can it, which took about two hours to cook down on the stove to thicken up, in addition to the processing time in the canner, putting heat and humidity in the kitchen. Now I use dried tomatoes for the spaghetti sauce and only can things that can be done in a shorter time. We can have the driest weather, but when the tomato harvest is at its peak, the clouds roll in to put the damper on solar drying. On the sunny days I dry tomatoes in my solar food dryers and on the cloudy days I can them. One good thing about canning is that you can process a lot of food in a relatively short period of time. Although I could use the water bath canner for the tomatoes, I use the pressure canner because the processing time is faster and it takes less water.

I never liked to stand over a stove while dipping tomatoes in boiling water to get the skins off. Instead of canning stewed, skinless tomatoes, I juice them. I cut them up and put them through a Foley food mill or the Victorio Strainer that I have. These mills separate the juice from the skins and seeds. The Foley mill is handiest to use and fits over a three quart pan or bowl. However, the tomatoes need to be cooked some first. The Victorio Strainer that I bought in 1986 clamps to the table and handles raw, cut-up tomatoes. I also use it to juice grapes. With the Victorio, there is no cooking required until the jars of tomato juice go into the canner. Tomato juice can be used as the base for soup with the addition of leftovers or dried vegetables. Once the tomatoes are juiced I can use the juice to make tomato soup,[6] adding butter, onions, and parsley or celery before canning.

If you want to make thick spaghetti sauce, instead of cooking it so long on the stove you could put the juice in the refrigerator overnight. The solids will sink and you can pour off the tomato water, canning it separately as soup stock. Use the rest for your sauce and it won't take so long to cook down thick. There are other brands of food mills on the market besides the ones I mentioned. You might even be able to find one at a yard sale.

Even if you have a place to store canning jars, you have the canners and other equipment to contend with. In our first house, we had a tiny kitchen. Not only did I have a water bath canner and a pressure canner, I had a four-gallon stainless steel pot that I prized. We put a shelf over our kitchen door and put them there. Where we live now, we have high ceilings and all my large pots sit above the kitchen cabinets. If you have space between your kitchen cabinets and the ceiling make the best use of it. I happen to be tall and have long arms, so I appreciate high storage. If necessary, keep a fold-up stool handy to reach up there. My Foley food mill hangs with my saucepans in full view in my kitchen. The Victorio Strainer and all its parts are stored in its original box on a shelf in the pantry.

Solar Food Dryers

For many years, I had an electric nine-tray Excalibur dehydrator that I bought used from a friend. I rarely used it because I didn't have time to learn something new, and because it pumped hot, humid air into the kitchen and made noise. Since I was already canning, which involved a lot of heat and humidity, it was probably the noise that bothered me the most. I couldn't put the dehydrator in another room because there was no other room available. Now, with the children grown and only two of us at the dinner table, I discovered that I preferred making spaghetti sauce from dried tomatoes rather than fresh. I no longer needed quarts of sauce. I could have canned it in pints, but with dried tomatoes, I could make any amount I wanted without spending hours in a hot kitchen in the summer cooking and canning sauce. I control the thickness of the sauce by how much water is added. For spaghetti sauce, I usually add water at twice the amount of dried tomatoes. I'm trying various paste tomatoes to see which I prefer to work with. So far, Principe Borghese is my favorite because it matures in 60 days (although the seed catalogs list it as 78 days), giving me a head start on the tomato drying season. The tomatoes in the solar driers might take more than a day to dry, but I can leave them drying overnight.

Home-canned spaghetti sauce is a wonderful convenience food. When I made it to preserve in large batches, I added basil, onions, garlic, and sweet peppers—all from the garden. Now that I make it from dried tomatoes I still have those ingredients available. I grow enough garlic to last from one season to another, so that is added fresh. I dry the basil and peppers easily enough. The onions I added to the sauce I was canning were ones that I suspected wouldn't store well. Those are the ones I dry now, so I have dried onions to put in sauce made from dried tomatoes. Recognize that creativity reigns in your kitchen and you can make up recipes as you go along. I can put just about anything in the sauce from dried ingredients. Dried zucchini and okra come to mind, which would also be good if I were making a sauce from tomato juice, since they are good thickeners.

With an increasing interest in dried food, I also had more interest in learning to dry it with the sun. I still didn't enjoy the electric dehydrator and I wanted to lessen my use of fossil fuel. Drying things on screens in the sun, and bringing them in the house overnight to go back out the next day, was not going to work for me. I read *The Solar Food Dryer* by Eben Fodor and wondered if that would work here in humid Virginia. I made his SunWorks (SW) design and it worked! The SW dryer has 60 percent of the drying capacity as the Excalibur that I had. Wanting to dry more produce at one time, I could have made another dryer just like the first one, but I wanted to explore other designs. I found plans online from *Home Power* magazine—issue numbers 57 and 69—for a solar dryer designed at Appalachian State University (ASU). There is a photo of both dryers in the color section of this book. The ASU dryer has 2.25 times the drying capacity as the SW dryer and 1.35 times the capacity of the nine tray Excalibur. I used an old storm window for the glazing on the SW dryer. That design could be made bigger by using a larger window, but then it would be harder to move around. I leave the ASU dryer out all winter, but the SW dryer is stored in the barn since it is easily moved.

With the solar dryers sitting in the garden and the sun shining, I am always looking around for something to put in them. I have dried tomatoes, peppers, peaches, apples, okra, zucchini, onions, collards, kale, and grapes for raisins. I've also dried snap beans, but didn't like them enough to use on a regular basis, so I'll stick with canning and salting for those. I dry parsley and celery leaves in the dryers, but hang other herbs, such as basil, sage, and thyme in the kitchen to dry. The collards and kale were successful; however, since they're available fresh through the winter in the garden, I find I rarely use the dried ones. I have met one woman who dries all sorts of greens to put in the smoothies she makes every day.

Early in my canning years, I thought it was great that I could buy apples from an orchard in the fall and can enough applesauce for the year. Now, I think that's too much work and too much fossil fuel. I dry apples in the solar dryers and store them in glass jars. I make the best-tasting applesauce as we need it throughout the year with dried apples.

It cost me about $120 to make the SW dryer with wood framed screens. Adding the electric option was another $23. I find I don't use that, preferring to go with solar power only. The ASU dryer cost me $385 to make. You can find more details about these dryers on my blog at HomeplaceEarth.wordpress.com. Once I began to use the solar dryers in earnest, my old Excalibur stopped working. We tried to fix it, but to no avail. That certainly freed up a spot in the kitchen.

Grain Mills

Grain mills are included in this chapter because they are necessary kitchen equipment to grind the grains you will be producing and storing. My first mill was a Corona mill, which is actually designed as a corn mill. It is not something I would want to use regularly for grinding wheat for bread, but it gave us an opportunity to make cornmeal and crack wheat to make hot cereal. It was an affordable starting point and you have to start somewhere. When I became more serious about grinding wheat for flour in 1999 I bought a Country Living Mill and was not disappointed. I mounted it on a piece of ¾ inch plywood and clamped that to the kitchen counter with C-clamps. That way I could move it if I wanted and I didn't have to drill holes in the counter.

In 2010 I had an opportunity to buy a GrainMaker mill. Remembering what a huge decision it was when I bought the Country Living Mill, I thought this would be a good opportunity to compare the two. I tested the mills side-by-side[7] and found that for the same effort (number of revolutions of the handle) the GrainMaker mill delivered twice the flour as the Country Living Mill. The fineness of the flour from both mills was the same. If the Country Living Mill had the extended handle attached, the effort to turn the handle was the same as the GrainMaker, otherwise the GrainMaker was easier. The GrainMaker sits on my kitchen counter now. The Country Living Mill has gone to live with one of our grown children. There are many grain mills on the market. If at all possible, try one before buying it. If not, carefully read the descriptions of folks who have used them and written about it. If a mill is too hard to use, it won't be used regularly and is of little use if you are trying to grow a sustainable diet.

Once grain is ground it begins to lose nutrients. Being able to store grains whole and grinding them as you need the flour allows you to offer your family healthier meals. These grain mills are not cheap, no doubt about it, but you should consider them as part of your health care. Your meals will be more nutritious, keeping you healthier; and you will be getting exercise, also keeping you healthier. If you are considering the cost of a good grain mill, check out the prices people pay for exercise equipment and gym memberships. Your grain mill purchase will prove to be a bargain in many ways.

These methods of storing and preserving are quite different from what I did 30 years ago. It is exciting for me to have evolved in this way and to have food from the garden that needs little equipment and fuss before we eat it. My knowledge, skills, and confidence needed to expand before I could reach this point. I hope that I've helped you to arrive at a similar place a little faster.

12

Sheds, Fences, and Other Stuff

WHEN YOU GARDEN, you have some tools you use and you have to keep them somewhere. An old (or new) mailbox, mounted near the garden can hold trowels. When our mailbox needed to be replaced, we mounted the old one in the garden. The door had fallen off already, making it easy for me to grab a trowel as I walk by. There are other things besides trowels that I use, and for those things, I need to dedicate space somewhere. For me, all these years, it has been a corner of the garage. A shed would be much better and it is actually in the plans for this summer.

Garden Shed

Okay, so now you know that, as I write this, I don't have a shed in the garden. I put one on my garden map because our time is opening up this summer to actually build it. As I said, my garden tools are in the corner of the garage, I hang the garlic, onions, peanuts, corn, and bags of beans in the barn, and I put other garden things in the goat shed (no goats there). One shed for everything, in the garden, would be ideal. Once the garden shed is built, it will free up space in these other places, although I'll probably still hang the corn, peanuts, and bags of beans in the barn until shelled and threshed.

If you are planning a shed, like we are, you need to assess what will go in it. Get out your graph paper and map the footprint and each wall. Gather everything that you will need to put there and measure it. Will it fit? Having defined my garden methods over the years, I know what tools I need. I use a garden spade (flat on the end), a garden fork (thick flat tines), and a mattock for digging chores. Sometimes I'll use the regular shovel, but we use that for jobs other than in the garden, so that might stay in the garage. My long handled tools include a four-tined cultivator and a hoe. The trowel assortment contains a regular trowel, a soil knife and a Trake (trowel on one end, cultivator/rake on the other). They seem happy living in the mailbox, so that's where they'll stay. I have a CobraHead tool that I really like. I will be hanging that in the new shed, along with my Japanese sickle. The machete is stored in a cloth sheath on a belt that will hang in the shed, also. You will find photos of these tools in the color section of this book.

In addition to the tools, I will have a bucket or two to hold soil amendments, wooden flats, a watering wand, and some extra plastic pots. Then there is a homemade six foot measuring stick, my one hundred foot tape measure, a sharpening stone for the sickle, and twine. One side should have a shelf—maybe two—with space underneath for the buckets. And I'll need a nail to hang a clipboard.

I want to post my garden map somewhere and I think the inside of the door would be a great place. When the door is open, I will be able to see the map and anything else I post there. Of course, a sliding door is nice and takes less space to open. I think I'll opt for a regular door though. I can also hang some of the small tools there, such as the sickle and CobraHead. Open the door and there they are.

I want the shed to be tall enough to hang the garlic and onions from the rafters to dry. If there is room above the rafters, I could put boards down on part of it, for a loft space providing extra storage. Currently I have a post in the garden that I hang my scale from when I'm weighing things. I want to extend an arm on the outside of the shed to hang the scale from. I'll have to make sure it's not somewhere that will hit me in the head when I walk by. The scale will need a spot inside the shed, also. A shelf to put my notepad on when I'm weighing things would be handy. I have a large galvanized tub that I'll probably hang on the outside of the shed. If the eaves extend out some, they will provide some protection from the weather for whatever I hang on the outside.

One reason we haven't built this shed yet (besides being busy with life in general) is because I hadn't refined what I wanted. If your garden supplies and tools are stored in multiple places like mine, chalk it up to still refining your plan. I didn't want to put up a plywood structure, so that took some thought. The siding will be 1" × 6" boards. The old chicken house and goat shed on our property have oak board siding. I could get oak fence boards from a local sawmill to do the job, but we'll be getting the lumber from daughter Betsy and her husband, Chris. They have a portable sawmill and can cut whatever size boards we want. The type of wood will be what's available. Their machine is a Wood-Mizer. If you were interested in buying lumber from a sawmill such as theirs, you could contact the Wood-Mizer company to find owners in your area.

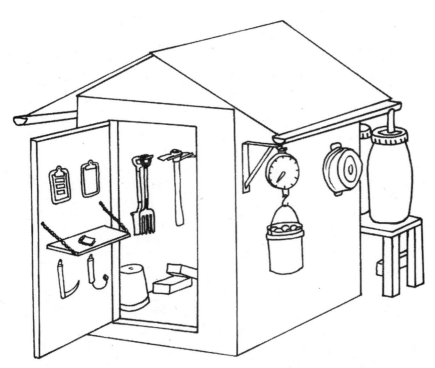

Even if they bought the sawmill for their own use, folks like this would be happy to sell some lumber to help pay for their mill. There are other brands of sawmills and I'm sure you could contact the owners in the same manner.

Water Storage

Wherever you have a roof is an opportunity to collect water. The easiest way to collect water is to put buckets or an animal water trough at the drip line of your building. You could also put gutters on the building and direct the water to one central place, either right at the building or a distance away. If rains are few and far between, one barrel of water won't go far, but it's handy to have to dip your watercan in for watering things in your cold frame. (Add watercan to the list for the shed contents.) For more storage space you need many barrels or a large holding tank. You can buy them new at farm supply stores or find them advertised used. Make sure whatever you get for storing water has not held anything toxic to your garden.

Outdoor Washing Station

I might not have had a garden shed, but I've had an outdoor washing station since I began selling vegetables in 1992. If you are cleaning vegetables for the markets, you will want to wash them somewhere other than in your kitchen. Washing your produce in your garden keeps the water and the soil there. Make a way to collect the water so you can return the wash water to the garden. Whatever soil washes off the produce also makes its way back to the garden in the process.

The equipment that lasted me for more than the ten years that I sold vegetables was an old bathtub set on cement blocks. I could fit a five-gallon bucket under the drain hole to catch the water, switching it out with another when it filled. I made a frame of 2 × 4s covered with half-inch hardware cloth (also known as rat wire) that fit the top of the tub and was used to drain produce on. Every few years I had to replace the wooden frame, but reused the same wire each time.

Beside that set-up I had built a bench out of scrap wood. It was at the right height to put the buckets of water on that I swished the lettuce in, before putting the lettuce on the wash table to be sprayed with the hose. There are drinking water safe hoses available. Make sure you use them for water that will be used with produce destined for someone's dinner table. That washing station was located in a spot that received morning shade, which is the time that I would be washing produce. If it was in the sun, I would have needed to have a roof.

I had some large plastic trays like the ones the bread companies use when delivering bread. I could pick tomatoes and peppers in those, set them on my washing screen, and spray with a hose. The tub caught the water, which drained into the buckets. If I had something small like cherry tomatoes or snap beans, I added a piece of quarter-inch hardware cloth, sized to fit inside the tray. That way none of the small vegetables fell through. Once drained, the produce could be packed for the market, with minimal handling in the washing process. If I hadn't had those

plastic bread trays, I would have made picking trays from wooden 2 × 4s and half-inch hardware cloth.

The washing station I have now contains a stainless steel free-standing sink that I found at a yard sale years ago. I clamped a hose onto the pipe going to one of the faucets so I actually have running water there—when the hose is turned on. A bucket by the drain catches the water. I don't have as large a quantity of produce to clean at one time for our kitchen as I did for the market, but I still think it is a valuable part of the garden. I can wash and cut vegetables for the solar food dryers, filling the dryer trays right in the garden. Since it is not in the shade, this spot needs a roof. We have made a structure out of bamboo, which is fun, but it will eventually be replaced with a permanent metal roof, like the one I'll put on the garden shed, with gutters to direct the rainwater. I have put down bricks as a floor to define the space. Long range plans for under this roof include an earth oven.

Coldframe

My coldframes are designated places to start seeds for transplant production and used all year long, not just when protection from the weather is needed. We harvest lettuce for the table from there from the fall until time to plant the seeds for the early spring crops. Our hardier winter-harvested greens are under low tunnels in the garden. You can build coldframes from just about anything. All you need is some type of surround to hold the glazing, which could be a piece of clear plastic, an old window, or a specially built top. The sides need to be high enough to provide growing room, but not so high that they shade the inside. I recommend you read *Four Season Harvest* by Eliot Coleman for a better understanding of season extension. *The 12-Month Gardener* by Jeff Ashton is also a good read to get your creative juices flowing for designing your season extension structures. Trust your instincts and use what is available. If you are growing on a large scale and have many structures, you will want to have more uniformity of size to make your work easier, but use what you have to get started. If what you build doesn't perform to your expectations, study it so you can learn for the next time.

Chicken House

I'm not going to tell you how to build a chicken house, but there are a lot of books out there that will. I will, however, tell you the best tips that I followed for my chicken house way back when. *Gene Logsdon's Practical Skills* was published in 1985, the year after we bought our five acre farm. Logsdon suggested making it possible to divide the chicken's living area into two, allowing you to separate your flock. I did and have been pleased with the arrangement ever since. The top of the divider is made from 2" × 4" fencing. The bottom of the divider is solid, but can be removed. It is connected to a post on each side with one nail at each place—these are easily pounded out if I want to open up the space. Once my chicks come out of the brooder (an old chicken tractor with a heat lamp) at about a month old, they go into the smaller of the two areas made by that divider. The chicks in this pen have their own run that has smaller wire than the rest of the outside enclosure. All the chickens, young and old, can see each other. When the young ones are ready to join the flock, I take out the divider. Logsdon's plan shows a larger chicken house with a door between the spaces.

We already had an existing building for a chicken house, or rather, three sides of one. I took that opportunity to build an overhang on that fourth side, since it needed work anyway, and put the rabbit cages there. The chickens could run under those cages on the outside of the building. I store the feed inside the chicken house, which is a lot better than walking to the barn to get it. There are two doors, covered with chicken wire, that separate the feed area from the two chicken areas. I found the doors at a yard sale, just at the time I needed them. If you keep your eyes and mind open and your intentions clear, things will show up for you like that. I built nest boxes that extend into the "feed room" to make egg collection easier. They were made from scrap wood with the tongues of old tennis shoes as hinges.

Although I couldn't find this in Logsdon's book when I checked, I'm sure I got the idea to put a loft in the chicken house from something he'd written. That building isn't tall, but I put some plywood over the chicken area and it makes a great place to store straw from the grain harvest or grass clippings for the chickens, bringing those birds into the circle of production. I put the straw down as bedding and feed them the dried "grass hay." When it is time to clean the chicken house, all of that becomes compost, which will go to the garden beds when finished. I cover the chicken droppings with hay or straw throughout the year, cleaning it out only once in the summer. It is important to have plenty of carbon material at all times to add to the chicken droppings. We have an old five foot wooden ladder that I keep folded up in the chicken house. I use that to reach the material in the loft.

Trellises

You can increase space in your garden by using trellises to get things off the ground. The garden maps in Chapter 8 show a trellis being used with tomatoes and cucumbers. When I started gardening, I used whatever I could find to hold things up. I don't like to put up strings to support plants, so I avoid that when possible. What I find most useful as a trellis is a piece of metal fence supported by metal posts that can be moved to the next bed in the rotation each year. Although I have used some homemade tomato cages, I prefer the fence for tomatoes.

Fencing

There are all types of fencing to choose from, depending on what you want to fence in or out. Notice the gauge of the wire on the fencing you find—the higher the number (gauge) the thinner the wire. It might make a difference when you are making comparisons. There are a number of plastic options, often with solar electric chargers involved. You'll have to look elsewhere for information on those systems and a good place to look is Premier One Supplies.[1] The plastic will need disposed of sooner than wire fencing, unless you compare it to chicken wire, which will just rust away. The batteries on the solar chargers will eventually need to be replaced, another disposal problem. We used electric fence (solar charger and aluminum wire) to expand the grazing area when we had the cow. I much prefer the permanent fencing that we have installed as we were able.

Here are some things I've experienced over the years with fencing:

Chicken wire will keep chickens in, but not necessarily keep other animals out. If you are having trouble with rabbits in your garden, a quick and relatively inexpensive fix is to put up 2' high chicken wire. You can staple it to wood posts, but if you use metal fence posts (the kind for electric fence) you could take it down easier. The metal fence posts have insulators for the electric wire that can adjust to different levels on the post. The chicken wire can hook right to those insulators. You can step over the fence to get in and out. This fence will be effective for maybe two years against rabbits, then they will just hop over it. You could go with a taller fence from the get-go, but you would have to put in a gate. In five years, the chicken wire will be rusting and the grass will be growing up in it. It is hard to trim close to chicken wire with a weed whacker, if that is what you are doing. If you plant a border around the outside, that would keep the grass away. A 2' high chicken wire fence is what I used as my first fence.

Welded wire fencing is much stronger than chicken wire. If there is a standard garden fence, I would have to say it is 4' tall welded wire with 2" × 4" spaces. Rabbits won't be jumping over this fence, but baby rabbits can squeeze through the 2 × 4 openings. You can fix that by attaching 2' tall chicken wire against the bottom. The way I would do that is by using twist ties, the kind used with plastic bags. I hope you have been saving them. The welded wire fence will keep out dogs and groundhogs, as long as they don't burrow under. You could dig a trench with your mattock where you want the fence and bury the wire at least four inches when you put it up. Because of the baby rabbit problem, I use 4' tall welded wire with 1" × 2" spacing for my fence.

Livestock panels are heavy 4 gauge wire fence panels that are 50" tall and 16' long. For support they need a metal T-post at each end and one in the middle. They're sometimes known as cattle panels. Some panels have a closer spacing at the bottom, but otherwise, the spaces between the galvanized rods (4 gauge wires) are 6" × 8". They would keep large animals out, but if you wanted to fence out small critters, you would need to put something else with it on the lower part. If you had a large dog in the yard, this might be the fence for you. It will keep the dog out of the garden and the dog will keep the rabbits and groundhogs away.

These panels are relatively easy to put up and take down, which could be useful if you haven't decided on permanent boundary lines for your garden or pasture. They can be cut with a hack saw or with strong bolt cutters. Livestock panels have many other uses besides fencing. They make a great trellis for tomatoes and cucumbers and are flexible enough to bend to make an arch. I have just such an arch going into my garden that grapevines grow over. Some people make a series of arches with them to make a greenhouse, covering it with heavy plastic. Hog panels are livestock panels that are 34" tall and have smaller spaces at the bottom.

Woven wire fence is what you see most often in pastures. It needs strong corner posts with brace posts about 6 feet away on each side. Unlike welded wire, woven wire fence can be pulled tight (stretched) and needs to be nailed with fence staples to wooden posts at the corners and at least every 100 feet. The line posts in between can be metal T-posts. The spaces between the wires are 6" × 6". The corner posts need to be strong 6" posts. The brace posts can be 4". Set the posts at least 2½' in the ground, 3' if possible for the corner posts. You might see Class 1 woven wire fence, but keep looking until you find Class 3 galvanized. It will last much longer.

There is sheep and goat woven wire fence that has 4" × 4" spaces. If they have horns, sheep and goats may get their heads caught in regular woven wire fence. We were doing some fencing with the 4" woven wire and decided to fence our barnyard while we were at it, inspired by the pigs we once had, as I mentioned in Chapter 10. Fencing the barnyard would make that grass available if the pasture was low and it is a layer of protection if we were loading animals in a truck or trailer. Not that we do that often, but we have in the past and it's nice to know that if an animal jumps out of the truck, it won't run over to the neighbor's place or into the road. The added bonus to fencing the barnyard is that the full grown chickens can't get through the 4" spacing. I hadn't anticipated that. They hop right through the 6" woven wire. Since their pen opens into the barnyard, I can let them out and they have access to pretty much everywhere except our yard and gardens. If you used woven wire fence around your garden, you would need smaller fencing to go with it, unless you only need to keep out large animals. You can find information about other types and sizes of metal fence and fencing for rotational grazing at Kencove Farm Fence Supplies.[2]

Board fencing could be used for a garden, but if you are keeping out small animals, you would be using a lot of boards. If it was a picket fence you would want the spaces between the pickets to be only an inch apart. Unless you have a long lasting wood, it tends to rot when in contact with the ground. So, although a picket fence looks great, it would rot from the ground up, letting the little critters slip underneath. Although we have used pressure treated posts in the corners of our pasture fence, for lack of other suitable posts at the time, I stay away from pressure treated wood in my garden because of the chemicals in them. We used round cedar posts for the wood posts in the garden.

Some of the pasture fence we installed in 1986 is made of four rows of oak boards 1" × 6" × 16' that we bought from a sawmill. The boards are spaced 6" apart, making the fence 4' high. It keeps in the livestock, but other wildlife, dogs, and chickens slip through it. The same year we put in our fence, someone down the road was putting in a similar fence, except they used pressure treated boards, the kind you would find in

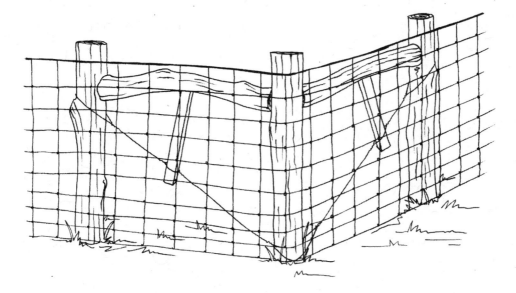

a building supply store. They were using the same kind of posts as we were, but they weren't set as deep in the ground. In not too many years, the boards started to warp and pull away from the posts. Then the posts failed, since they weren't set deep enough. That fence has long been taken down, but ours only needs occasional repair. I have to admit, it could use a paint job, but structurally, it's in good shape. It will pay you dividends in the long run to pay attention to details, and spend a little more, if you want a long lasting fence.

Black locust is something we decided we needed to learn more about so we could avoid using pressure treated posts. I have read accounts of black locust fence posts being dug up after forty years and used again somewhere else. We found the seedlings available from the Virginia Department of Forestry and planted some on the east side of our barn-yard. We were not familiar with how they grew and didn't want to cause a problem with the neighbor's hayfield to the north of us. Black locust will send out new trees several feet away. We planted them where they could be controlled by mowing by us and the neighbor across the fence. Black locust trees have thorns that you need to watch out for, but they are a wonderful tree to grow for the bees who love the flowers. When the trees are cut, you can let multiple stems grow back from each stump for future cuttings. Growing black locust is a great permaculture element on our very small farm. Black locust provides long lasting wood for fence posts, gives us a nice eastern border to our property, is food for the bees, and fixes nitrogen in the ground, contributing to the health of the soil.

13

Rethink Everything!

By this time I hope you have so many thoughts and ideas floating around in your head that you need time to process it all. Good—that shows that you are thinking, which is just what I want you to do. If I have left you wishing I would have written more about certain things, look up the resources I've suggested. I learned from so many people, many of them mentoring me from the words in their books. Others I have had a chance to meet and get to know. I've suggested a lot of books and I know people have limited budgets, so look to your library for some of these. That's how I got started. If the book you want is not on the shelf, inquire about getting it through an interlibrary loan or ask the library to buy it. That's what libraries do. If you never ask for it, they won't know there is an interest. Making the best use of the library will also make those books available to others. If you find a book you would like to have in your personal library, buy it from the author if you can. It benefits the author and you've made a connection. Some books may be out of print, but used copies are easy to find with the wonders of the internet.

Planning a sustainable diet is a different way of looking at both eating and growing. Now you understand what a sustainable diet is—eating in a way that replenishes the earth—and how to work out your own plan.

Making maps of your garden and your whole property (the perma-culture map) helps you discern what you have to work with. Post them somewhere so that you can keep them in your mind. Show them to others and talk about them. Get your friends involved.

Try new crops. Since I began studying Biointensive methods, my garden has grown to include grains, compost crops, sweet potatoes, cowpeas and other dried beans, greens (collards, kale, chard) grown all winter, and more Irish potatoes. I still grow tomatoes, peppers, lettuce, snap beans, herbs, and other things that add variety and pleasure to our meals, but the emphasis is on the staple crops and feeding back the soil. Attending my first Ecology Action Three-Day Workshop led by John Jeavons was what made me look at everything I was eating with new eyes. One example, and it doesn't have to do with staple crops, is that when I was growing up I drank tea with milk and sugar. That's just what you did. In my college dorm room, I had tea and sugar available, but no milk, so I began drinking tea with only sugar. Over the years, I changed to using honey instead of sugar. At that workshop in October 2000, I learned that it takes the life production of twelve bees to produce one teaspoon of honey![1] I stopped putting honey in my tea. I still sweeten things, but at a more sustainable rate, I hope. Tea itself has taken on a new meaning. I do like black tea, but there are so many things I can grow in my garden that I can also brew into tea. Usually these are the plants that make great companions to the other crops and to the bees. If you need examples, try bee balm, lemon balm, spearmint, sage, and hibiscus. And so it went with the rest of what my family was consuming. Our meals began to change. If we are eating our way through the earth's resources at an unsustainable rate, we affect everyone and everything on the planet.

I've given you some worksheets to use. It is hard to start with a blank slate. Please don't let these limit your efforts. You are learning to partici-pate in the dance of life. Think of these worksheets as the marks on the floor and the directions for folks learning a new dance. At first they put their feet on the marks and follow the directions. Once they understand how it all goes, and put their heart and soul into it, their dancing changes into something wonderful and completely their own. That's what I want for you. I want you to join in the dance and make it wonderful.

You could dance alone, but it is a whole lot more fun dancing with others. We are part of communities. You can live miles from anyone, but you are still part of it. Remember the permaculture ethics I mentioned in Chapter 1—care for the people, care for the earth, and redistribute the surplus? Some of that surplus is knowledge. Share what you have learned with others. For more information about permaculture, read *The Permaculture Handbook* by Peter Bane. This book gives excellent information about what permaculture is and how to make it a part of your life.

Gandhi said you have to be the change you want to see in the world. Live like you want the world to be and it will become that. It might take longer than you think it should, but it will happen. Meanwhile, you will be living the life you want. I began teaching at the community college as a result of a phone call to the horticulture department suggesting they offer a class in organic gardening. I had recognized a need that I thought I could fill. My classes developed into a whole program, and a Career Studies Certificate in Sustainable Agriculture was proposed in 2003. Bureaucracy is a slow moving machine, but in 2013 that certificate finally became a reality! Meanwhile, we didn't stand still waiting for it to happen all those years. I kept teaching the classes and my students went out to make their mark in the world in all areas of the food system. They are the movers and shakers in their communities, with or without a Career Studies Certificate. The college library recognized the interest in sustainable agriculture—that's how I know you can make a difference by visiting the library and taking out books—and as a result, the library at the Goochland (Virginia) Campus of J. Sargeant Reynolds Community College has the best collection of sustainable agriculture and permaculture books of anywhere I know.

As I mentioned in the introduction, I felt a need to leave the college in order to address a larger community. I started a blog—which was an adventure—and I wrote this book, which has been an even bigger adventure. Meanwhile, our daughter Betsy took over teaching the classes. She has put her heart and soul into that dance and made it her own, just as it should be. Having the library so well stocked with the books she and her students needed wasn't enough for her. She read an article in *Acres U.S.A.* magazine[2] about seed libraries and was so intrigued, she wanted to start one at JSRCC. It only took a year from "Wouldn't this be great?" to "The first orientation meeting for the JSRCC Seed Library is March 5, 2013." The wheels of bureaucracy often might seem stuck, but once you loosen the glue that has held them in place for so long, they begin rolling at a much faster pace.

Seed libraries are places where you can "borrow" seeds to plant, grow them out, and save seeds to return at the end of the season. Just by participating, you can help to develop strains of things unique to your area, keep little known varieties from becoming extinct, increase your seed saving knowledge, and get free seeds! Getting the library stocked with books and seeds is great, but it is only a success if others participate. We need people to dance with us. You don't have to start a seed library to make a difference. Giving a neighbor some of your garlic to plant, or taking some of your homegrown food to a potluck and engaging in conversation about how you grew it, can be enough to make a difference.

Eating a sustainable diet becomes a way of life. It is not just what comes from the garden, but every food choice. How was it grown? What resources were used to get it to your table? These are the questions you will be asking about everything you eat. Real living is when everything you do becomes your way of life. The boundaries begin to blur, or at least how you think of them begins to blur, as you go about your job, do laundry, maintain your home, cook dinner, and interact with your friends and family. Real living is when you begin to see everything as part of one joyful whole. Sometimes the change needs only to be made in how you think of things. One example of that kind of change is if you have thought of weeding as drudgery, something you have to endure. Begin to think of weeding as a harvest of materials for the compost pile. Harvesting is a positive thing. You will be cleaning up your garden and building compost at the same time. When you change your thoughts, everything around you changes. Learn to look at everything in a new way.

We are living in exciting times. People around the world are waking up to what has been happening environmentally. Our survival depends on the wise use of our resources. Sure, we are running out of petroleum, but a petroleum based society is not the only way to live. Much has been written about riding a bike or taking the bus, not using disposable products, and recycling containers. Even if you do all that, it is not enough.

You may still be looking at things from the outside. You need to know that you are an integral part of it all. Every bite you take determines how you want the world to be used, in order to grow your food. I've given you the tools you need to plan and grow a sustainable diet. When you put yourself into your plan, hopefully you come to realize that, as I mentioned in Chapter 5, you are not separate from the earth that grows your food. Think of becoming part of your garden and not just the steward. When you do that, your garden becomes a sacred place.

Your approach to things might be different than mine. You'll come up with ideas that may or may not work, but you won't know for sure until you try them and get it out of your system. You don't need my approval for what you do in your garden. I do hope, however, that you consider the sustainability of all you do. When you take the time to consider that, it will direct your actions more than anything I could tell you. Pay attention to everything that comes your way. That is mindfulness. When you interact with people, take the time to look them in the eye and talk to them, rather than being distracted by your cell phone. When you are mindful of your actions and your surroundings, they have more meaning for you. Find a spot in your garden or in nature somewhere, quiet your mind, and sit for at least twenty minutes. Observe what is happening around you and become part of it. Feel the air and listen to the sounds. Going barefoot at least some of the time will "ground" you. You will be more aware of the soil temperature and the moisture in the ground. I like to be in my garden when the sun goes down and calmness settles in with the dew.

Recognize the interconnectedness of everything. Scientists used to think they had to take things apart to study them, and they imagined that if they knew everything about each piece they would know what they had. Now they are finding that they need to study the relationship of all those parts. They need to study the whole. The whole is greater than the sum of its parts. Acting together in communities, we are much stronger than we can ever be going it alone. A sustainable diet recognizes the whole and is part of the dance of life where everything is connected. Come join the dance and grow a sustainable diet!

Endnotes

Chapter 1
1. patternliteracy.com/419-fear-and-the-three-day-food-supply-3.
2. Rosemary Morrow, *Earth Users Guide to Permaculture*, 2nd ed. (Australia: Simon & Schuster, 2006), p. 9.

Chapter 3
1. John Jeavons, *How to Grow More Vegetables and Fruits, Nuts, Berries, Grains, and Other Crops Than You Ever Thought Possible on Less Land Than You Can Imagine.* 8th ed. (Berkeley, CA: Ten Speed Press, 2012).
2. Sandor Elix Katz, *The Art of Fermentation* (White River Junction, VT: Chelsea Green, 2012), p. 25.
3. Weston A. Price, *Nutrition and Physical Degeneration* (New York, London: Paul B. Hoeber Inc., 1939), Chapter 15, journeytoforever.org /farm_library/price/price15.html.
4. Timothy Johns, "Detoxification Function of Geophagy and Domestica-tion of the Potato," *Journal of Chemical Ecology*, Vol.12 No.3 (1986): 635–636.
5. "Sweet Potatoes Rank #1 in Nutrition by CSPI", sweetpotatoblessings .com/number1.htm.
6. International Institute of Tropical Agriculture (ITTA), iita.org/cassava.
7. Jeavons, *How to Grow More Vegetables*, pp.129–179.
8. At plantmaps.com you will find interactive USDA Plant Hardiness Zone Maps for the US, Canada, and the UK that will help you deter-mine which climate zone you are in.
9. Sally Fallon, *Nourishing Traditions*, rev.2nd ed. (Washington, DC: New Trends Publishing, Inc. 2001), p. 62.

Chapter 4
1. GrainMaker Mill. Bitterroot Tool & Machine, P.O. Box 130, Stevensville, MT 59870. Toll free (855) 777-7096. grainmaker.com.

2. *Booklet #30: GROW BIOINTENSIVE^SM Sustainable Mini-Farming Certification Program for Teachers and Soil Test Stations* is available as a free download at growbiointensive.org/publications_main.html.

Chapter 5

1. Mark Schonbeck, "Cover Cropping: On-farm, Solar-powered Soil Building," Virginia Association for Biological Farming Information Sheet, Number 1-06, 5/16/06. p.2.

Chapter 8

1. John Seymour, *The New Self-Sufficient Gardener* (New York: DK Publishing, 2008), pp. 215–216.

Chapter 9

1. savingourseeds.org. Saving Our Seeds is a venue for raising awareness of our threatened genetic resources, and a resource for providing information and knowledge tools for seed saving and seed conservation.

Chapter 10

1. *National Nutrient Database for Standard Reference, Release 25*, ndb.nal.usda.gov/ndb/search/list. Find information on all the nutrition resources mentioned in the Resources section of this book.
2. Cheryl Long and Tabitha Alterman, "Meet Real Free-Range Eggs." *Mother Earth News*, October/November 2007. motherearthnews.com/Real-Food/2007-10-01/Tests-Reveal-Healthier-Eggs.aspx#axzz2PR1jMRh7.
3. eatwild.com/nutritionnews/nutritionnews5.htm.
4. Louis M. Hurd, *Modern Poultry Farming*, 4th ed. (New York: The Macmillan Company, 1956), p. 143.
5. Joel Salatin, *Pastured Poultry Profits*, Polyface Inc., 1996. p. 256.
6. David B. Weems, *Raising Goats: The Backyard Dairy Alternative* (Blue Ridge Summit, PA: Tab Books Inc., 1983), p. 184.
7. Livestock panels are 16' long rigid fence panels with spaces about 6" × 8". They can be secured by three metal T-posts or tied to trees.
8. Fernando Funes, Luis García, Marin Bourque, Nilda Pérez, and Peter Rosset, *Sustainable Agriculture and Resistance* (Oakland, CA: Food First Books, 2002), pp. 6–7, 158–159.
9. polyfacefarms.com/2011/07/25/forage-based-rabbits/.
10. Helga Olkowski, Bill Olkowski, Tom Javits, and Farallones staff, *The Integral Urban House* (Canada: New Catalyst Books, 2008), p. 268.
11. Jeavons, *How to Grow More Vegetables*, p. 155.

Chapter 11

1. Roger B. Yepsen, Jr., ed., *Home Food Systems* (Emmaus, PA: Rodale Press, 1981), p. 242–243.
2. Klaus Kaufmann and Annelies Schöneck, *Making Sauerkraut and Pickled Vegetables at Home* (Vancouver, Canada: Alive Books, 2002), pp. 33–41.
3. *USDA Complete Guide to Home Canning* download link nchfp.uga.edu /publications/publications_usda.html
4. *USDA Complete Guide to Home Canning*, 196 pages. Order through https://mdc.itap.purdue.edu/item.asp?item_number=AIG-539 or call toll free (888)-398-4636.
5. Website link for fact sheets and for ordering *So Easy to Preserve*, 5th ed. nchfp.uga.edu/publications/publications_uga.html.
6. You will find the recipe for tomato soup to can on the recipe page of my blog at homeplaceearth.wordpress.com/recipes/.
7. homeplaceearth.wordpress.com/2011/05/03/grain-mill-comparison -country-living-vs-grainmaker.

Chapter 12

1. Premier One Supplies. Portable fencing, solar chargers, and more. premier1supplies.com.
2. Kencove Farm Fence Supplies. All types of fencing. kencove.com.

Chapter 13

1. Honey Bee Facts from the American Beekeeping Federation. abfnet.org /displaycommon.cfm?an=1&subarticlenbr=71.

Resources

Homeplace Earth

HomeplaceEarth.wordpress.com is Cindy's blog. Consider it free continuing education for growing a sustainable diet. Cindy's website is Home placeEarth.com.

Chapter 1. Sustainable Diet

Some organizations that help the world's struggling people lead more sustainable lives:

Ecology Action teaches people worldwide to better feed themselves while building and preserving the soil and conserving resources. growbio intensive.org.

Heifer International works with communities to end hunger and care for the earth. heifer.org.

Lambi Fund of Haiti works toward economic justice, democracy, and alternative sustainable development in Haiti. lambifund.org.

Trees, Water, & People develops and manages continuing reforestation, watershed protection, renewable energy, appropriate technology, and environmental education programs in Latin America and the American West. treeswaterpeople.org.

Chapter 2. Garden Maps

Morrow, Rosemary. *Earth Users Guide to Permaculture*, 2nd ed., Australia: Simon & Schuster, 2006. Distributed by Chelsea Green in the US.

Smyser, Carol A. *Nature's Design*. Emmaus, PA: Rodale Press, 1982. Helpful map-making information.

Chapter 3. Crop Choices

Bountiful Gardens, 1712-D S. Main St., Willits, CA 95490. bountifulgardens .org. Besides seeds, Bountiful Gardens is a source for Ecology Action GROW BIOINTENSIVE® Sustainable Mini-Farming publications.

Deppe, Carol. *The Resilient Gardener*. White River Junction, VT: Chelsea Green, 2012.

Jeavons, John. *How to Grow More Vegetables and Fruits, Nuts, Berries, Grains, and Other Crops Than You Ever Thought Possible on Less Land Than You Can Imagine*, 8th ed. Berkeley, CA: Ten Speed Press, 2012.

Seymour, John. *The New Self-Sufficient Gardener*. New York: DK Publishing, 2008.

Southern Exposure Seed Exchange, P.O. Box 460, Mineral, VA 23117. southernexposure.com.

Wallace, Ira. *The Timber Press Guide to Vegetable Gardening in the Southeast*. Portland, OR: Timber Press, 2013.

Weston A. Price Foundation. westonaprice.org.

Chapter 4. How Much to Grow

Allen, John. *Bioshphere 2: The Human Experiment*. New York: Penguin Books, 1991.

GrainMaker Mill. Bitterroot Tool & Machine, P.O. Box 130, Stevensville, MT 59870. Toll free (855) 777-7096. grainmaker.com.

Conner, Cindy. "A Plan for Food Self-Sufficiency," *Mother Earth News*, October/November 2012, p. 32. motherearthnews.com/homesteading-and-livestock/food-self-sufficiency-zmoz12onzkon.aspx#axzz2cRq4yzlf.

Conner. "Filberts In My Garden." homeplaceearth.wordpress.com/2012/03/20/hazelnuts-filberts-in-my-garden/.

Conner. "Homegrown Fridays," "Homegrown Fridays 2012," and "Homegrown Fridays 2013." homeplaceearth.wordpress.com/category/homegrown-fridays/.

Conner. "Using a Piteba Oil Press." homeplaceearth.wordpress.com/2012/07/10/using-a-piteba-oil-press/.

Jeavons. *How to Grow More Vegetables*. Master Charts in Chapter 8.

Silverstone, Sally. *Eating In: From the Field to the Kitchen in Biosphere 2*. Oracle, AZ: The Biosphere Press, 1993.

Tucker, Sherry Leverich. "Making Sorghum," Mother Earth News (blog). goo.gl/g3JXw.

Chapter 5. Cover Crops and Compost—
Planning for Sustainability

Dawling, Pam. *Sustainable Market Farming*. Canada: New Society, 2013.

Jenkins, Joseph. *The Humanure Handbook*, 3rd. ed. PA: Jenkins Publishing, 2005. To order call toll free (814)786-9085 or order online at humanurehandbook.com. This website is full of information, including videos.

Schonbeck, Mark. "Cover Cropping: On-farm, Solar-powered Soil Build-
ing," Virginia Association for Biological Farming Information Sheet,
Number 1-06, 5/16/06. sare.org/Learning-Center/Project-Products
/Southern-SARE-Project-Products/Cover-Cropping-On-Farm-Solar
-Powered-Soil-Building.
Sustainable Agriculture Network. *Managing Cover Crops Profitably*, 3rd ed.
2007. To order copies call (301)779-1007, email sanpubs@sare.org, or
order online at sare.org.

Chapter 6. Companion Planting

Benson, Brinkley, Dr. Richard McDonald, Dr. Ronald Morse, *Farmscaping
Techniques for Managing Insect Pests*. drmcbug.com/EPM/EPM3/Farm
scaping%20FinalVT.doc.
Cunningham, Sally Jean. *Great Garden Companions*. Emmaus, PA: Rodale
Press, 1998.
Dufour, Rex. *Farmscaping to Enhance Biological Control*. 2000. https://attra
.ncat.org/attra-pub/summaries/summary.php?pub=145.
Riotte, Louise. *Carrots Love Tomatoes: Secrets of Companion Planting for
Successful Gardening*, 2nd ed. Storey Publishing, 1998.
Symbiont Biological Pest Management Company, Dr. Richard McDonald.
drmcbug.com.
Yepsen, Roger B. Jr., ed. *Organic Plant Protection*. Emmaus, PA: Rodale
Press, 1976.

Chapter 7. Plan for Food When You Want It

Climate information for your region including interactive plant, tree, and
gardening maps and data for the U.S., Canada, and UK. plantmaps
.com.

Chapter 8. Rotations and Sample Garden Maps

Coleman, Eliot. *The New Organic Grower*, 2nd ed. White River Junction,
VT: Chelsea Green, 1996. An 8-year crop rotation is explained in Chap-
ter 7—Crop Rotation.
Dawling. *Sustainable Market Farming*. A 10-year Crop Rotation Pinwheel
is explained in Chapter 4—Crop Rotations for Vegetables and Cover
Crops.
Katz, Sandor, *Wild Fermentation*. White River Junction, VT: Chelsea
Green, 2003.
Rodale, J. I., ed. *How to Grow Vegetables & Fruits by the Organic Method*.
Emmaus, PA: Rodale Press, 1961.

Seymour, *The New Self-Sufficient Gardener*. A 4-year crop rotation is explained in Chapter 3—Planning the Food-Producing Garden.

Chapter 9. Seeds

Ashworth, Suzanne. *Seed to Seed*, 2nd ed. Decorah, IA: Seed Savers Exchange, 2002.

Dawling. *Sustainable Market Farming*. Chapter 56—Seed Growing, and Chapter 57—The Business of Seed Crops.

Jeff McCormack is the founder and previous owner of Southern Exposure Seed Exchange and Garden Medicinals and Culinaries, co-founder of Virginia Plant Savers, and founder and owner of McCormack's Botanicals, and JHM Designs. He is also the lead author of *Bush Medicine of the Bahamas*.

Navazio, John. *The Organic Seed Grower*. White River Junction, VT: Chelsea Green, 2012.

Saving Our Seeds is a venue for raising awareness of our threatened genetic resources, and a resource for providing information and knowledge tools for seed saving and seed conservation. savingourseeds.org.

Seed Companies I've found helpful. All sell cover crop seeds and all have signed the Safe Seed Pledge:

Bountiful Gardens, 1712-D S. Main St., Willits, CA 95490-4411. (707) 459-6410. bountifulgardens.org. Heirloom and open-pollinated seeds for sustainable growing.

Fedco Seeds, P.O. Box 520, Waterville, ME 04903-0520. fedcoseeds.com. Fedco is a cooperative with a very informative catalog. Call (207) 426-0090 to request a catalog. No phone orders.

High Mowing Seeds, 76 Quarry Rd, Wolcott, VT 05680. (802) 472-6174. highmowingseeds.com. Offers only 100% certified organic seeds.

Johnny's Selected Seeds, P.O. Box 299, Waterville, ME 04903. Toll free (877) 564-6697. johnnyseeds.com. An employee-owned company. Catalog contains equipment for market growers.

Southern Exposure Seed Exchange, P.O. Box 460, Mineral, VA 23117. (540) 894-9480. southernexposure.com. A worker-run cooperative. Catalog carries almost all open-pollinated seeds and seed saving supplies.

Sow True Seed, 146 Church St., Asheville, NC 28801. (828) 254-0708. sowtrueseed.com. Specializing in heirloom, certified organic, and traditional Southern Appalachia varieties.

Territorial Seed Company, 20 Palmer Ave, Cottage Grove, OR 97424. Toll free (800) 626-0866. territorialseed.com. Publishes a winter catalog in July. Season extension supplies.

Chapter 10. Including Animals

Dietary Reference Intakes (DRI) tables iom.edu/Activities/Nutrition /SummaryDRIs/~/media/Files/Activity%20Files/Nutrition/DRIs/5 _Summary%20Table%20Tables%201-4.pdf.

Fallon, Sally. *Nourishing Traditions*, rev. 2nd ed. Washington, DC: New-Trends Publishing, 2001.

Kirschmann, John D. and Nutrition Search, Inc. *Nutrition Almanac*, 6th ed. New York: McGraw-Hill, 2007.

Lee, Andy and Patricia Foreman. *Chicken Tractor*, 3rd ed. Good Earth Publications, 2011.

Long, Cheryl and Tabitha Alterman. "Meet Real Free-Range Eggs." *Mother Earth News*, October/November 2007. motherearthnews.com/Real -Food/2007-10-01/Tests-Reveal-Healthier-Eggs.aspx#axzz2PR1jMRh7.

National Nutrient Database for Standard Reference, Release 25. ndb.nal.usda .gov/ndb/search/list.

Olkowski, Helga, Bill Olkowski, Tom Javits and the Farallones Institute Staff. *The Integral Urban House.* Canada: New Catalyst Books. This book was first published in 1979 by the Farallones Institute and republished in 2008 by New Catalyst Books.

Pennington, Jean A. T. and Judith Spungen. *Bowes & Church's Food Values of Portions Commonly Used*, 19th ed. New York: Lippincott Williams & Wilkins, 2010.

Salatin, Joel. *Pastured Poultry Profits.* Swoope, VA: Polyface Inc., 1996.

Ussery, Harvey. *The Small-Scale Poultry Flock.* White River Junction, VT: Chelsea Green, 2011.

Weems, David B. *Raising Goats: The Backyard Dairy Alternative.* Blue Ridge Summit, PA: Tab Books, 1983.

eatwild.com. Information on the benefits of raising animals on pasture and a link to local farms selling grassfed products.

Chapter 11. Food Storage and Preservation

Bubel, Nancy and Mike. *Root Cellaring.* Emmaus, PA: Rodale Press, 1979. This book was republished in 1991 by Storey Publishing.

Fallon. *Nourishing Traditions*, rev. 2nd ed. This book gives directions for making fermented vegetables by the jar.

Georgia Cooperative Extension Service. *So Easy to Preserve*, 5th ed. Website link to order: setp.uga.edu. As of August 2013, this 375 page book was $18, including shipping. For more information call (706) 542-2657 or write Office of Communications, 117 Hoke Smith Annex, Cooperative Extension Service, The University of Georgia, Athens, GA 30602-1456.

Lehman's. This business has supplies and equipment for living with less electricity, including European style crocks (with water trough). Visit their store On the Square in Kidron, Ohio. P.O. Box 270, Kidron, OH 44636. Toll free (888) 438-5346. lehmans.com.

Kaufmann, Klaus and Annelies Schöneck. *Making Sauerkraut and Pickled Vegetables at Home.* Canada: Alive Books, 2002.

Katz. *Wild Fermentation.* Also by Katz is *The Art of Fermentation*, Chelsea Green, 2012. More information at wildfermentation.com.

Mother Earth News Fairs are held at Puyallup, Washington in June, Seven Springs, Pennsylvania in September, and Lawrence, Kansas in October. Find out more at motherearthnews.com/fair.

National Center for Home Food Preservation website: nchfp.uga.edu/.

USDA Complete Guide to Home Canning. Download link: nchfp.uga.edu /publications/publications_usda.html. Print version: 196 pages. As of August 2013 the price was $18 plus shipping. Order at https://mdc.itap .purdue.edu/item.asp?item_number=AIG-539 or call toll free (888)- 398-4636.

Stoner, Carol and the staff of Organic Gardening and Farming. *Stocking Up.* Emmaus, PA: Rodale Press, 1973. There have been some newer editions published.

Yepsen, Roger B. Jr., ed. *Home Food Systems.* Emmaus, PA: Rodale Press, 1981.

Dehydrators—Electric and Solar

Excaliber electric dehydrators. 3355 Enterprise Ave., Suite 160, Fort Lauderdale, FL 33331. Toll free (800) 875-4254. excaliburdehydrator .com.

Scanlin, Dennis, Marcus Renner, David Domermuth, and Heath Moody. "Improving Solar Food Dryers". *Home Power* #69 (February/March 1999). wot.utwente.nl/nl/wp-content/uploads/improving-solar-food -dryers.pdf.

Scanlin, Dennis. "Indirect, Through-Pass, Solar Food Dryer." *Home Power* #57 (February/March 1997) homepower.com/view/?file=HP57_pg62 _Scanlin.

Fodor, Eben. *The Solar Food Dryer.* Canada: New Society, 2006.

Homeplace Earth blog posts about solar food dryers: homeplaceearth.word press.com/category/food-preservation/solar-food-dryers.

Grain Mills

My first mill was a Corona King Convertible with a tall hopper and both stone and steel grinding plates. I found I mostly used the steel plates. The Corona Grain Mill is listed in many home brewing catalogs.

Country Living Grain Mill. Country Living Products, 14727 56th Ave. NW, Stanwood, WA 98292. (360) 652-0671. countrylivinggrainmills.com.

GrainMaker Mill. Bitterroot Tool & Machine, P.O. Box 130, Stevensville, MT 59870. Toll free (855) 777-7096. grainmaker.com.

Homeplace Earth blog post comparing grain mills: homeplaceearth.word press.com/2011/05/03/grain-mill-comparison-country-living-vs-grain maker.

Chapter 12. Sheds, Fences, and Other Stuff

Ashton, Jeff. *The 12-Month Gardener.* New York: Lark Books, 2001.

CobraHead tools. Online store. Toll free (866) 962-6272. cobraheadllc.com.

Coleman, Eliot. *Four Season Harvest*, rev. ed. White River Junction, VT: Chelsea Green, 1999. Coleman's *The Winter Harvest Handbook*, published by Chelsea Green in 2009 has information on using unheated greenhouses for year-round vegetable production. *The Four Season Farm Gardener's Cookbook* by Eliot Coleman and Barbara Damrosch, published by Workman Publishing in 2012, contains plans and ideas for making and using a 10' × 12' portable greenhouse.

Kencove Farm Fence Supplies. All types of fencing. 344 Kendall Rd., Blairsville, PA 15717. There are also warehouses in Indiana and Missouri. Toll free (800) 536-2683. kencove.com.

Logsdon, Gene. *Gene Logsdon's Practical Skills.* Emmaus, PA: Rodale Press, 1985.

Premier One Supplies. Portable fencing, solar chargers, and more. 2031 300th Street, Washington, IA 52353. Toll free (800) 282-6631. premier1 supplies.com.

Purple Mountain Organics. Professional gardening tools. Online Store. Toll free (877) 538-9901. purpletools.net.

Way Cool Tools. Professional gardening tools. Online store. P.O. Box 235, White Hall, VA 22987. Toll free (877) 353-7783. waycooltools.com.

Chapter 13. Rethink Everything!

Bane, Peter. *The Permaculture Handbook.* Canada: New Society, 2012.

Hibiscus tea. Homeplace Earth blog post on growing, drying, and using Red Thai Roselle Hibiscus tea. homeplaceearth.wordpress.com/2013/02 /05/red-thai-roselle.

J. Sargeant Reynolds Community College, 1851 Dickinson Rd., Goochland, VA 23063. reynolds.edu.

McDorman, Bill and Stephen Thomas. "Sowing Revolution: Seed Libraries Offer Hope for Freedom of Food." *Acres U.S.A.*, January 2012.

Seed library at the JSRCC Goochland Campus, Virginia. jsrcclibrary.word press.com/2013/03/14/goochland-library-hosts-seed-library.

Index

About the Author

Cindy Conner grew up in Parkman, Ohio. Having been active in 4-H, she naturally went to Ohio State University where she met her husband and received a degree in Home Economics Education. Since she learned to make her own clothes in 4-H, sewing was her first area of interest in home economics. With arms and legs longer than average, she realized you don't have to settle for what's available in the store when you can make your own clothes to fit.

Being a stay-at-home mom to their four children gave Cindy the opportunity to carry that kind of thinking over to the food she prepared for her family. Her study of organic gardening and sustainability led her to become a market gardener, then an educator.

Cindy Conner founded Homeplace Earth to provide permaculture education with an emphasis on sustainable food production, including getting the food to the table using the least fossil fuel. She researches how to sustainably grow a complete diet in a small space at her home near Ashland, Virginia, and has produced the videos *Develop a Sustainable Vegetable Garden Plan* and *Cover Crops and Compost Crops IN Your Garden*. Cindy was instrumental in establishing the sustainable agriculture program at J. Sargeant Reynolds Community College in Goochland, Virginia and taught there from 1999–2010. You can follow Cindy's blog at HomeplaceEarth.wordpress.com.

All the worksheets in this book are available
for download at http://tinyurl.com/mf4a33r

If you have enjoyed *Grow a Sustainable Diet*, you might also enjoy other

BOOKS TO BUILD A NEW SOCIETY

Our books provide positive solutions for people who
want to make a difference. We specialize in:

**Sustainable Living ◆ Green Building ◆ Peak Oil
Renewable Energy ◆ Environment & Economy
Natural Building & Appropriate Technology
Progressive Leadership ◆ Resistance and Community
Educational & Parenting Resources**

New Society Publishers

ENVIRONMENTAL BENEFITS STATEMENT

New Society Publishers has chosen to produce this book on recycled paper made
with 100% post consumer waste, processed chlorine free, and old growth free.

For every 5,000 books printed, New Society saves the following resources:[1]

28	Trees
2,495	Pounds of Solid Waste
2,745	Gallons of Water
3,580	Kilowatt Hours of Electricity
4,535	Pounds of Greenhouse Gases
20	Pounds of HAPs, VOCs, and AOX Combined
7	Cubic Yards of Landfill Space

[1]Environmental benefits are calculated based on research done by the Environmental Defense Fund and
other members of the Paper Task Force who study the environmental impacts of the paper industry.

For a full list of NSP's titles, please call 1-800-567-6772 or check out our web site at:

www.newsociety.com